Mohammad Ahsanullah
and Mohammad Shakil

A Compendium of Univariate Statistical Distributions

www.novapublishers.com

DOI: https://doi.org/10.52305/MRTA0751

Library of Congress Cataloging-in-Publication Data

ISBN: 979-8-89530-727-4 (Softcover)
ISBN: 979-8-89530-904-9 (eBook)

Published by Nova Science Publishers, Inc. New York

To my family – Mohammad Ahsanullah
To my wife, children, late parents and teachers – Mohammad Shakil

Contents

Preface

Statistical distributions are the bedrock of statistical theory and practice, offering models that describe the behavior of data across countless domains, from engineering to economics, biology to social sciences. In particular, univariate statistical distributions, those that describe the distribution of a single variable, are of immense importance. They help statisticians model randomness, quantify uncertainty, and make inferences about populations from sample data. With this book, *A Compendium of Univariate Statistical Distributions*, we aim to provide a comprehensive, accessible, and systematic guide to the fundamental univariate distributions encountered in statistical analysis.

The journey of this book begins with an exploration of a variety of univariate distributions, ranging from the widely known and studied distributions such as the Normal, Exponential, and Binomial, to the less commonly encountered yet equally important ones like the Birnbaum-Sanders, and von Mises distributions. Each chapter is dedicated to a particular distribution, providing the reader with a deep dive into its characteristics, and basic properties, which are essential for their practical applications and further research on statistical distributions.

The structure of the chapters has been designed with clarity and accessibility in mind. Each distribution is introduced with a detailed discussion on its foundational aspects, followed by an examination of its basic properties, such as moments, skewness, kurtosis, and other statistical features. Additionally, the practical aspects of random number generation for each distribution are provided, equipping the reader with the tools necessary for simulation and computational work.

A particular strength of this compendium is its wide range of distributions. From the simplicity of the Uniform and Geometric distributions to the complexity of distributions like the Lévy and Burr, this book strives to be a thorough reference for anyone interested in univariate statistical analysis. Whether you are a student embarking on your statistical journey, a researcher seeking a deeper understanding of the distributional models available, or a practitioner in need of a reliable source for applied statistics, this book will serve as an invaluable resource.

Throughout this book, our goal has been not only to provide detailed theoretical insights but also to offer practical guidance on the use of each distribution in applied settings. We hope that readers will find this work not only informative but also inspiring, helping to deepen their understanding of the power and versatility of statistical distributions in the world of data analysis.

M. Ahsanullah
Professor Emeritus
Rider University, Lawrenceville, NJ, USA
E-mail: ahsan@rider.edu

and

M. Shakil
Professor
Miami Dade College, Hialeah, FL, USA
E-mail: mshakil@mdc.edu

Acknowledgments

We extend our gratitude to all the statisticians and researchers whose work has shaped the field of statistical distributions, and to the readers who engage with this book. We trust that it will serve as a valuable reference for years to come. Furthermore, the first author, Mohammad Ahsanullah, and the second author, Mohammad Shakil, would like to express their gratitude to their respective institutions, Rider University and Miami Dade College, for all support and assistance received throughout the preparation and completion of this compendium. We would like to thank Nova Science Publishers for publishing the book.

Introduction

Statistical distributions are the bedrock of statistical theory and practice, offering models that describe the behavior of data across countless domains, from engineering to economics, biology to social sciences. In particular, univariate statistical distributions, those that describe the distribution of a single variable, are of immense importance. They help statisticians model randomness, quantify uncertainty, and make inferences about populations from sample data.

A Compendium of Univariate Statistical Distributions, we aim to provide a comprehensive, accessible, and systematic guide to the fundamental univariate distributions encountered in statistical analysis.

The journey of this book begins with an exploration of a variety of univariate distributions, ranging from the widely known and studied distributions such as the Normal, Exponential, and Binomial, to the less commonly encountered yet equally important ones like the Birnbaum-Sanders, and von Mises distributions. Each chapter is dedicated to a particular distribution, providing the reader with a deep dive into its characteristics, and basic properties.

About the Authors

Dr. Mohammad Ahsanullah is a professor Emeritus of Rider University, Lawrenceville, New Jersey, USA. He is a Fellow of American Statistical Association and of Royal Statistical Society. He is a member of International Statistical Institute. He has authored and co-authored more than fifty books and more than four hundred papers in reputable journals. His areas of research are Record Values, Order Statistics, Characterizations of Distributions, etc.

Dr. M. Shakil, Ph.D., CStat, PStat, is a Professor of Mathematics and Statistics, Miami Dade College, Hialeah Campus, Miami, Florida, USA. He has been teaching mathematics and statistics courses at Miami Dade College, for many years. He has co-authored four statistics textbooks and several research articles in reputable journals. His research interests are Distribution Theory, Characterizations, Statistical Inferences, etc. He is a member of many international learned societies.

Chapter 1

Preliminaries

1.1. Mathematical Functions and Constants

The modified Bessel function of first kind of integer order n, $I_n(x)$ is as follows

$$I_n(z) = \frac{1}{\pi}\int_0^{\pi} e^{zcos\theta}\cos(n\vartheta)d\theta$$

$I_n(z)$ can be expressed in the following power series

$$I_n(z) = (\frac{z}{2})^n \sum_{k=0}^{\infty} \frac{1}{\Gamma(k+1)\Gamma(k+n+1)}(\frac{z}{2})^{2k}.$$

$Fon\ n = 0, we\ obtain$

$$I_0(z) = \sum_{k=0}^{\infty} \frac{1}{(k!)^2}(\frac{z}{2})^{2k}$$

$$= 1 + \frac{z^2}{2^2} + \frac{z^4}{2^2 4^2} + \frac{z^6}{2^2 4^2 6^2} + \ldots$$

Modified Bessel function of second kind of integer order n, $K_n(x) is\ as$

$$K_n(x) = \int_0^{\infty} e^{-xcosh\,nx}\cosh nx\, dx$$

$$= \frac{\pi}{2\sin nx}(I_{-n}(x) - I_n(x))$$

$$K_n(x) = \frac{x^n}{2^n n!}(1 - \frac{x^2}{2.2n+2} + \frac{x^4}{2.2n+2.2n+4} - \cdots \ldots)$$

$\gamma(Euler's Constant)$

$$= \lim_{n\to\infty}(1 + \frac{1}{2} + \frac{1}{3} + \ldots + \frac{1}{n} - \ln n)$$

$$= \sim 0.5772$$

$$e = \lim_{n\to\infty}(1 + \frac{1}{n})^n$$

$$= \sim 2.7183$$

$$e^x = \lim_{n\to\infty}(1 + \frac{x}{n})^n$$

For x>0,

$\Gamma(x+1) = \int_0^\infty e^{-u} u^x du$
$= e^{-u}u^{-x}|_0^\infty + x\int_0^\infty e^{-u} u^{x-1} du$
$=0+ x\int_0^\infty e^{-u} u^{x-1} du$
$= x\,\Gamma(x)$
$=x(x-1)\,\Gamma(x-1)$
$\Gamma(1) = \int_0^\infty e^{-u}\, du =1.$

Thus if x is an integer

$\Gamma(x+1) = x!$
$\Gamma(\mathrm{m}) = \infty$, if m is 0 or negative.
$\Gamma(\mathrm{x})\Gamma(1-\mathrm{x}) = \frac{\pi}{sin\pi x}$

Incomplete gamma functions

$\Gamma(x,\alpha) = \int_\alpha^\infty u^{x-1} e^{-u} du$
$\gamma(x,\alpha) = \int_0^\alpha u^{x-1} e^{-u} du$

$$\mathrm{n}! = \Gamma(\mathrm{n}+1) = \int_0^\infty x^n e^{-x} dx$$

putting x =ny, we obtain

$$\mathrm{n}! = e^{nlnn}\, n \int_0^\infty e^{n(lnn-y)}\, dy$$

Using Laplace's approximation to integrals

$\int_0^\infty e^{n(lny-y)} dy \cong \sqrt{\frac{2\pi}{n}}\, e^{-n},$
we have
$\mathrm{n}! = \sqrt{(2\pi n)}(\frac{n}{e})^n$
$\Gamma(\frac{1}{2})=\sqrt{\pi}$

Catalan number

The n-th Catalan number C_n *is*

$$C_n = \frac{(2n)!}{n!(n+1)!} = \frac{4^n\Gamma(n+\frac{1}{2})}{\sqrt{\pi}\Gamma(n+2)} = \frac{1}{n}\binom{2n}{n-1}$$

$Digamma\ function\ , \Psi(x) = \frac{d}{dx} ln\Gamma(x) = \frac{\Gamma\prime(x)}{\Gamma(x)}$

$$\Psi(1) = -\gamma$$
$$\Psi(x) = -\gamma + \sum_{k=1}^{x-1} \frac{1}{k}$$

Tri gamma function, $\Psi'(x) = \frac{d^1}{dx} ln\Gamma(x) = \frac{d}{dx}\frac{\Gamma\prime(x)}{\Gamma(x)}$

$$\Psi'(x) = \sum_{k=0}^{\infty} \frac{1}{(x+n)^2}$$
$$\Psi'(x+1) = \Psi'(x) - \frac{1`}{x^2}$$
$$\Gamma(a)\Gamma(b) = \int_0^\infty e^{-u}u^{a-1}\, du \int_0^\infty e^{-v}v^{b-1}\, dv$$
$$= \int_0^\infty \int_0^\infty e^{-u-v}\ u^{a-1}\ v^{b-1}\ \mathrm{dudv}$$

Put u =st and v-s(1-t), then

$$\Gamma(a)\,\Gamma(b)$$
$$= \int_{s=0}^\infty \int_{t=0}^1 e^{-s}\,(s)^{a-1}(s(1-t)^{b-1}sdtds$$
$$= \int_0^\infty e^{-s}\, s^{a+b-1}\mathrm{ds} \int_0^1 t^{a-1}(1-t)^{b-1}\mathrm{dt}$$
$$= \Gamma(a+b)B(a,b)$$

Thus

$$B(a,b) = \frac{\Gamma(a)\Gamma(b)}{\Gamma(a+b)}$$

$Incomplte\ beta\ function$

$$\mathrm{B(x,a,b)} = \int_0^x u^{a-1}(1-u)^{b-1}du$$

$Pochhammer\ symbol$

$$(a)_n = \mathrm{a(a+1)\,(a+2)\ldots\ldots(a+n-1)}$$
$$(a)_0 = 1$$

Generalized hyperbolic function

$$p^{F_q}\,(a_1, a_2, \ldots .\, a_p, b_1, b_2, \ldots, b_{q:}z)$$
$$= \sum_{n=0}^{\infty} \frac{(a_1)_n\,(a_2)_{n \ldots}(a_p)_n}{(b_1)_n\,(b)_{n \ldots}(b)_q)_n} \frac{z^n}{n!}$$
$$\binom{n}{r} = \frac{n!}{r!(n-r)} = \binom{n}{n-r}$$

Double factorials

(2n+1)!!=1.3.5…(2n+1),
(2n)!!= 2.4.6…(2n),

for integer n≥ 1.

Error function (Erf)

$$\text{Erf(x)} = \frac{2}{\sqrt{\pi}} \int_0^x e^{-u^2}\,\text{du}$$

Complementary error function (Erfc)

$$\text{Erfc(x)= 1-Erf(x)} = \frac{2}{\sqrt{\pi}} \int_x^{\infty} e^{-u^2}\,\text{du}$$

Bernoulli numbers,$B_n, n \geq 0$, are the coefficients of $\frac{x^n}{n}$ in the expansion of x in $\frac{x}{e^x-1}$

We have

$$\frac{x}{e^{x-1}} = \sum_{n=0}^{\infty} B_n \frac{x^n}{n}$$

$$B_0 - 1, B_1 = -\frac{1}{2} - B_2 = \frac{1}{6}. B_4 = -\frac{1}{30}, B_{2n+1} = 0, \text{n>0.}$$

1.2. Distributions

Sample Space (Ω): The set of all possible outcomes of an experiment.

Sigma algebra (σ): It is a collection of subsets of a sample space including Ω that satisfies the following three properties.

1) Closure under complementation. It says that if a set A is in sigma *algebra*, then $A^c$is also in sigma algebra.
2) closure under countable union. If $A_1, A_2, \ldots$ are in sigma algebra, then $U_{i=1}^{\infty} A_i$ must be in the sigma algebra.
3) Contain the sample sapce Ω.

Probability (p): It is a measure of the likelihood of an event to occur. Many events cannot be predicted with total certainty. Probability can range from 0 to 1, where 0 means the event to be an impossible one and 1 indicates a certain event.

Probability Space: (Ω, σ, P): *It consists* the sample space, the sigma algebra and the probability.

Random Variable: It is variable whose value is a function that assigns values to each of the experimenter's outcomes. A random variable can be discreet or continuous.

Discrete random variable: It takes countable number of distinct values. Suppose we toss a fair coin twice and X is the number of heads that occur in the two tosses. Then X takes the values 0,1 or 2 depending on the occurrences of the heads. Here X is a discrete random variable.

Continuous random variable: It can represent any value within a specific range or interval of the possible outcome of an experiment. An example is the measurement of a city's average monthly rainfall.

Cumulative Distribution Function (cdf). The cumulative distribution function of a real valued random variable X is the probability that X will take a value less and equal to x. It is denoted as F(x).

Probability Mass Function (PMF). It gives the probability for individual outcome for discrete random variable.

Probability density function (pdf): A non-negative continuous function f(x) related to the outcomes of an experiment which has the property $\int_{-\infty}^{\infty} f(x)dx = 1$ is known as the probability density function. It is the derivative of the cumulative distribution function F(x) with respect to x. For every interval A (a,b], the probability P(A) of A is

P(A) = $\int_a^b f(x)dx$.

Mathematical Expectation (E(X))

Let $x_1, x_2,,,,. x_n$ denote thg possible values of a discrete random variable X and let $p_1, p_2, \ldots, p_n$denote the corresponding probabilities. Then the mathematical expectation, E(X) is

$$E(X) = \sum_1^n x_i p_i.$$

If the random variable X is continuous with pdf f(x), then E(X) is

$$E(x) = \int_{-\infty}^{\infty} x f(x) dx.$$

Variance (Var(X))

The variance of a random variable X is the expected value of the squared deviation from the expected value of X, E(X) =μ.

$$\begin{aligned} Var(X) &= E(X - E(X))^2 \\ &= E(X^2 - 2XE(X) + (E(X))^2) \\ &= E(X^2) - 2(E(X))^2 + (E(X))^2 \\ &= E(X^2) - (E(X))^2 \end{aligned}$$

Let $x_1, x_2,,,,. x_n$ denote thg possible values of a discrete random variable X and let $p_1, p_2, \ldots, p_n$denote the corresponding probabilities. Then the Var(X) is

$$Var(x) = \sum_1^n x_i^2 p_i - (\sum_1^n x_i p_i)^2)$$

If the random variable X is continuous with pdf f(x), then Var(X) is

$$Varr(X) = -\int_{-\infty}^{\infty} x^2 f(x) dx - (\int_{-\infty}^{\infty} x f(x) dx)^2.$$

The r-th Moment (E(X^r)

Let $x_1, x_{2,,,,.}$ x_n denote thg possible values of a discrete random variable X and let $p_1, p_2, \ldots, p_n$denote the corresponding probabilities. Then the r-th moment (E(X^r) is

$$E(X^r) = \sum_1^n x_i^r p_i \text{ ,}$$

provided the summation is finite.

If the random variable X is continua with pdf f(x), then E(X) is

$$E(X^r) \text{ -} \int_{-\infty}^{\infty} x^r f(x) dx,$$

provided the integral is finite.

Median

The median is the middle value of a given set of data arranged in ascending or descending order.

Kurtosis (Kurt(X))

Kurt (X) is the fourth standardized moment , defined as

$$\text{Kurt}(X) = E(\frac{X-\mu}{\sigma})^4 = \frac{\mu_4}{\sigma^4}$$

Quartile

The first quartile (Q_1) is defined as the 25-th percentile where the lowest 25% of the data is below this point.

The second quartile (Q_2) is the median of a data set m thus 50% of the data lies below this point.

The third quartile (Q_3) is the 75th percentile point where lowest 75% of the data lies below this point.

Moment generating function (M(t))

Let $x_1, x_{2,,,,,} x_n$ denote thg possible values of a discrete random variable X and let $p_1, p_2, \ldots, p_n$denote the corresponding probabilities. Then Moment generating function (M(t)) is

$$M(t) - E(e^{tx})) = \sum_1^n e^{tx_i} p_i,$$

provided the summation is finite.

If the random variable X is continuous with pdf f(x), Then Moment generating function (M(t)) is

$$\text{M(t)} = \int_{-\infty}^{\infty} e^{tx} f(x) dx,$$

provided the integral is finite.

Characteristic Function ($\varphi(t)$

Let $x_1, x_{2,,,,,} x_n$ denote thg possible values of a discrete random variable X and let $p_1, p_2, \ldots, p_n$denote the corresponding probabilities. Then the characteristic function ($\varphi(t)$).is

$$\varphi(t))) = \sum_1^n e^{itx_i} p_i.$$

If the random variable X is continuous with pdf as f(x), then the characteristic function ($\varphi(t)$) is

$$\varphi(t) = \int_{-\infty}^{\infty} e^{itx} f(x) dx.$$

Hazard Rate (H(x))

Let $x_1, x_{2,,,,,} x_n$ denote thg possible values of a discrete random variable X and let $p_1, p_2, \ldots, p_n$denote the corresponding probabilities. Then the hazard rate H(x) is

$$\text{H}(x_i) = \frac{P(x_i)}{P(X \geq x_i)}. \text{ I=1.2,…,n.}$$

If the random variable X is continuous with pdf f(x), then the hazard rate H(x) is

$$H(x) = \frac{f(x)}{1-F(x)}.$$

1.3. Some Common Univariée Distributions

Arcsine

ARS(a,b)

PDF p(x) $\frac{1}{\sqrt{(x-a)(b-x)}}$

CDF F(x) $\frac{2}{\pi}\arcsin(\sqrt{(\frac{x-a}{b-a}}),$

$-\infty<a<x<b<\infty.$

Bernoulli

BER (p)

PMF $p(X=1) = 1- p(X=0) = p,\ 0<p<1$

CDF $F(x)= P(X\leq x) = 1-p,\ 0\leq x<1$

$=1,\ x\geq 1.$

Beta

BE(m,n)

PDF p(x) $\frac{1}{B(m,n)}\, x^{m-1}\,(1-x)^{n-1}$

CDF F(x) I_x (m,n)

$I_x(m,n) = \int_0^x \frac{1}{B(m,n)}\, u^{m-1}\,(1-u)^{n-1}du$

$B(m,n) = \frac{\Gamma(m)\Gamma(n)}{\Gamma(m+n)},$ m>0. n>0. 0<x<1.

Binomial

BN(n,p)

PMF $\quad p(x) = P(X=m) = \binom{n}{m} p^m (1-p)^{n-m}$

CDF $\quad F(x) = P(X \leq x) = \sum_{m=0}^{x} \binom{n}{m} p^m (1-p)^{n-m}$,

$0 < p < 1, n > 1$.

Birnbaum Sanders

BS ($\boldsymbol{\alpha}, \boldsymbol{\beta}, \boldsymbol{\lambda}$)

PDF $\quad f_X(x;\ \alpha,\ \beta,\ \lambda) \quad \frac{1}{2\alpha(x-\lambda)}\left[\sqrt{\frac{x-\lambda}{\beta}} + \sqrt{\frac{\beta}{x-\lambda}}\right] \times$

$\phi\left[\frac{1}{\alpha}\left\{\sqrt{\frac{x-\lambda}{\beta}} - \sqrt{\frac{\beta}{x-\lambda}}\right\}\right])$

CDF $\quad F_X(x;\ \alpha,\ \beta,\ \lambda) \quad \Phi\left[\frac{1}{\alpha}\left\{\sqrt{\frac{x-\lambda}{\beta}} - \sqrt{\frac{\beta}{x-\lambda}}\right\}\right]$,

$x > \lambda \geq 0,\ \alpha > 0,\ \beta > 0.$

$\phi(x) = \frac{1}{\sqrt{(2\pi)}} e^{-\frac{x^2}{2}}$ and $\Phi(x) = \frac{1}{\sqrt{(2\pi)}} \int_{-\infty}^{x} e^{-\frac{u^2}{2}}$ du.

Burr

BURR(3p) (k,α, β)

PDF $\quad$ f(x) $\quad k\alpha \frac{(\frac{x}{\beta})^{\alpha-1}}{\beta! 1 + (\frac{x-\gamma}{\beta})^{\alpha})^{k+1}}$

CDF $\quad$ F(x) $\quad$ 1- $(1 + (\frac{x}{\beta})^{\alpha})^{-k}$,

$k, \alpha, \beta > 0.$

Cauchy

CA (μ, σ)

PDF $\quad$ f(x) $\quad \frac{1}{\pi\sigma} \frac{1}{1+(\frac{x-\mu}{\sigma})^2}$

CDF F(x) $\frac{1}{2}+\frac{1}{\pi}\arctan(\sigma^{-1}(x-\mu))$,

$-\infty < \mu < x < \infty, \sigma > 0.$

Chi-Square

CHI(μ, σ, n)

PDF f(x) $\frac{1}{2^{\frac{n}{2}}\Gamma\left(\frac{n}{2}\right)} e^{-\frac{1}{2}(\frac{x-\mu}{\sigma})} (\frac{x-\mu}{\sigma})^{\frac{n}{2}-1}$

CDF F(x) $\gamma(\frac{n}{2}.\frac{x-\mu}{\sigma}$

$0 \leq \mu < x < \infty, \sigma > 0.$

$\gamma(x, y) = \int_0^y s^{x-1} e^{-s} ds.$

Discrete Uniform

DUN(n)

PMF P(x=x) $\frac{1}{n}$

CDF F(x) $\frac{x+1}{n}$,

n > 1,. x= 0,1,2,..,n-1.

Exponential

E(μ, σ)

PDF f(x) $\frac{1}{\sigma} e^{-(\frac{x-\mu}{\sigma})}$

CDF F(x) $1-e^{-(\frac{x-\mu}{\sigma})}$,

$-\infty < \mu < x < \infty, \sigma > 0.$

F

PDF $(\frac{m_1}{m_2})^{\frac{m_1}{2}} \frac{x^{\frac{m_1}{2}-1}}{B(\frac{m_1}{2},\frac{m_2}{2})} (1 + \frac{m_1}{m_2} x)^{-\frac{m_1+m_2}{2}}$

CDF $I(\frac{m_1 x}{m_1+m_2}, \frac{m_1}{2}, \frac{m_2}{2})$

$I(x,a,b) = \frac{B(x,a,b)}{B(a,b)}$,

$x>0.\ m_1 > 0, m_2 > 0.$

Gamma

GA(a,b)

PDF f(x) $\frac{1}{\Gamma(n)b^a} x^{a-1}\, e^{-x/b}$

CDF F(x) $\frac{1}{\Gamma(a)}\, \gamma\left(\frac{x}{b}, a\right)$,

$0 \le x < \infty, a > 0. b > 0.$

Geometric

GE (p)

PMF P(X=x) $p(1-p)^x$

CDF F(x) $1-(1-p)^{x+1}$,

x=0,1,2,…., 0<p<1.

Gumbel Distribution

$GU(\mu, \sigma)$

PDF f(x) $\frac{1}{\sigma} e^{-\frac{x-\mu}{\sigma}} e^{-e^{-\frac{x-\mu}{\sigma}}}$

CDF F(x) $e^{-e^{-\frac{x-\mu}{\sigma}}}$,

$-\infty < \mu < x < \infty,\ \sigma > 0.$

Hypergeometric

HG (a,b,n)

PMF P(X=x) $\frac{\binom{a}{m}\binom{b}{n-m}}{\binom{a+b}{n}}$

n=1,2,…,a+b, m = max((0,n-b)…min(n,a))

Inverse Gaussian

I G (μ, λ)

PDF f(x) $(\frac{\lambda}{2\pi x^3})^{1/2} e^{-\frac{\lambda}{2x}(\frac{x-\mu}{\mu})^2}$

CDF F(x) $\Phi(\sqrt{\frac{\lambda}{x}}\left(\frac{x}{\lambda}-1\right))+e^{2\frac{\lambda}{\mu}}\,\Phi(-\sqrt{\frac{\lambda}{x}}\left(\frac{x}{\lambda}+1\right))$,

$\Phi(\mathrm{x}) = \frac{1}{\sqrt{2\pi}} \int_{-\infty}^{x} e^{-\frac{u^2}{2}}\, \mathrm{du}$,

x>0, $\mu > 0, \lambda > 0$.

Laplace

LA (μ, σ)

PDF f(x) $\frac{1}{2\sigma} e^{-|\frac{x-\mu}{\sigma}|}$

CDF F(x) $\frac{1}{2}\, e^{-\frac{x-\mu}{\tau\sigma}}, if\ 0 < x < \mu,$

F(x) 1-$\frac{1}{2}\, e^{-\frac{x-\mu}{\sigma}}$ if $\mu < x < \infty, \sigma > 0$.

Levy

LV (μ, σ)PDF $\sqrt{\frac{1}{2\pi}} \frac{e^{\frac{\sigma}{2(x-\mu)}}}{(x-\mu)^{3/2}}$, x>$\mu$, $\sigma > 0$

CDF $2 - 2\Phi\left(\sqrt{\frac{\sigma}{x-\mu}}\right)$,

where

$$\Phi(x) = \int_{-\infty}^{x} \frac{1}{\sqrt{(2\pi)}}\, e^{-\frac{u^2}{2}}\, \mathrm{du}.$$

Logarithmic Series

LS (p)

PMF $\frac{-p^x}{x ln(1-p)}$, x=1.2., 0<p<1

CDF $1+\frac{B(p,x+1,0)}{\ln(1-p)}$.

Logistic

LO $((\mu,\sigma)$

PDF f(x) $\frac{1}{\sigma}\frac{e^{-\frac{x-\mu}{\sigma}}}{(1+e^{-\frac{x-\mu}{\sigma}})^2}$

CDF F(x) $\frac{1}{1+e^{-\frac{x-\mu}{\sigma}}}$,

$-\infty<\mu<x<\infty,\sigma>0.$

Lognormal

LN(μ,σ)

PDF f(x) $\frac{1}{\sigma x\sqrt{2\pi}}e^{-(\frac{lnx-\mu}{\sigma})^2}$

CDF F(x) $\frac{1}{2}[1+\text{erf}\,(\frac{lnx-\mu}{\sigma\sqrt{2}})]$,

erf(x) = $\frac{2}{\sqrt{\pi}}\int_{-\infty}^{x}e^{-\frac{t^2}{2}}$ dt

$0<x<\infty,-\infty<\mu<\infty,\sigma>0.$

Lomax

LO (α,σ)

PDF f(x) $\frac{\alpha}{\sigma}(1+\frac{x}{\sigma})^{-(\alpha+1)}$

CDF F(x) 1-$(1+\frac{x}{\sigma})^{-\alpha}$,

0$\leq x<\infty,\alpha>0.\sigma>0.$

Negative Binomial

NB (m,p)

PMF p(x=k) $\binom{k+m-1}{k}p^m(1-p)^k$, m>0

CDF F(X$\leq k$) $I(p,m,k+1)$,

k=0,1,2,…,, 0<p<1, m is an integer.

Normal

N (μ, σ)

PDF f(x) $\frac{1}{\sigma\sqrt{2\pi}} e^{-(\frac{x-\mu}{\sigma})^2}$

CDF F(x) $\frac{1}{2}$(1+erf $(\frac{x-\mu}{\sigma\sqrt{2}})$),

-∞ $< \mu < x <$∞,$\sigma > 0$.

Pareto

PA (σ, δ)

PDF f(x) $\delta \frac{\sigma^\delta}{x^{\sigma+1}}$

CDF F(x) 1-$(\frac{\sigma}{x})^\delta$,

0<$\delta < x < \infty, \sigma > 0$.

Poisson

P (λ)

PMF P(X=x) $\frac{e^{-\lambda}}{x!} \lambda^x$.

CDF P(X$\leq x$) $\sum_{k=0}^{x} \frac{e^{-\lambda}}{k!} \lambda^k$,

x=0,1.2,…, $\lambda > 0$.

Power Function

POW (a,b,δ)

PDF f(x) $\frac{\sigma}{b-a} (\frac{x-a}{b-a})^{\sigma-1}$

CDF F(x) $(\frac{x-a}{b-a})^\sigma$,

=∞ $< a < x < b < \infty, \sigma > 0$.

Rayleigh

RA (μ, σ)

PDF f(x) $\frac{x-\mu}{\sigma} e^{-\frac{1}{2}(\frac{x-\mu}{\sigma})^2}$

CDF F(x) $1- e^{-\frac{1}{2}(\frac{x-\mu}{\sigma})^2}$,

$0 \le \mu < x < \infty . \sigma > 0.$

Student's t

ST(n)

PDF f(x) $\frac{1}{\sqrt{n\, B(n/2, 1/2))}})(1+\frac{x^2}{n})^{-(n+1)/2}$

CDF F(x) $\frac{1}{2}$+x$\Gamma(\frac{n+1}{2})\frac{{}_3F_2(\frac{1}{2},\frac{n+1}{2},\frac{3}{3}|-\frac{x^2}{2})}{\sqrt{(\pi n)}\Gamma(\frac{n}{2})}$,

$-\infty < x < \infty$, n$\ge$ 1.

Triangular

TR (μ, σ)

PDF f(x) $\frac{x-\mu}{\sigma^2}$, if $\mu < x < \mu + \sigma$

$\frac{\mu+2\sigma-x}{\sigma^2}$, if $\mu + \sigma < x < \mu + 2\sigma$

CDF F(x) $\frac{(x-\mu)^2}{2\sigma^2}$, if $\mu < x < \mu + \sigma$

F(x) $1 - \frac{(\mu+2\sigma-x)^2}{2\sigma^2}, if\ \mu + \sigma < x < \mu + 2\sigma$

=1 if x$\ge \mu + 2\sigma$.

Uniform

UN (a,b)

PDF f(x) $\frac{1}{b-a}$,

CDF F(x) $\frac{x-a}{b-a}$,

$-\infty < a < b < \infty$.

Von Mises

VM (μ, σ)

Pdf $\frac{e^{\sigma \cos(x-\mu)}}{2\pi I_0(\sigma)}$

CDF $\frac{1}{2\pi}$(x+$\frac{2}{I_0(\sigma)} \sum_{j=0}^{\infty} I_j(l) \frac{sin j\frac{x-\mu}{\sigma})}{j\tau\sigma}$),

$\mu - \pi < x < \mu + \pi, -\infty < \mu < \infty$..

Weibull

WE (μ, σ, δ)

PDF f(x) $\frac{\delta}{\sigma}(\frac{x-\mu}{\sigma})^{\delta-1} e^{-(\frac{x-\mu}{\sigma})^{\delta}}$

CDF F(x) $e^{-(\frac{X-\mu}{\sigma})^{\delta}}$

-0≤ $\mu < x < \infty. \sigma > 0, \delta > 0.$

Chapter 2

Arcsine Distribution (ARS(a,b))

2.1. Introduction

pdf $\frac{1}{\pi}\frac{1}{\sqrt{(x-a)(b-x)}}$, $-\infty < a < b < \infty$

The pdf of ARS (0,1) is given in figure 2.1.

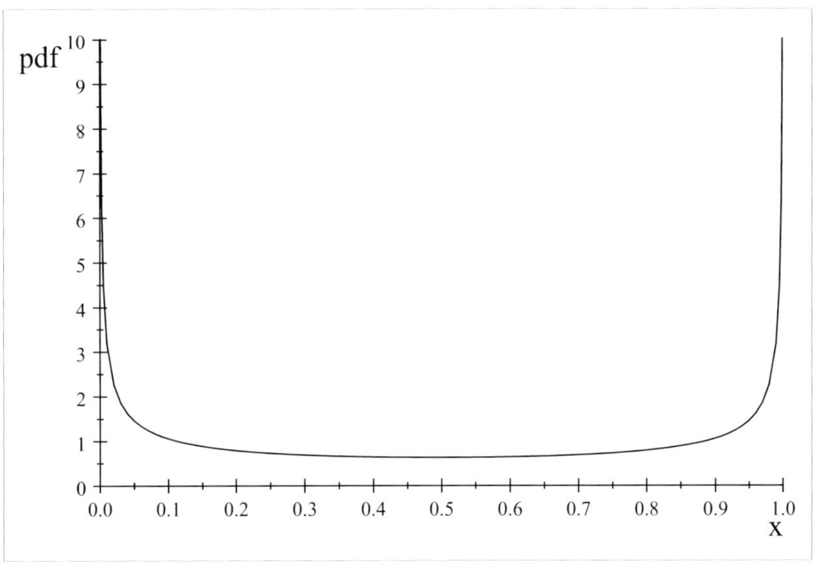

Figure 2.1. PDF ARS(0,1).

Cdf $\frac{2}{\pi}$ Arcsine($\sqrt{(\frac{x-a}{b-a}}$)

The cdf of ARS (0,1) is given in figure 2.2.

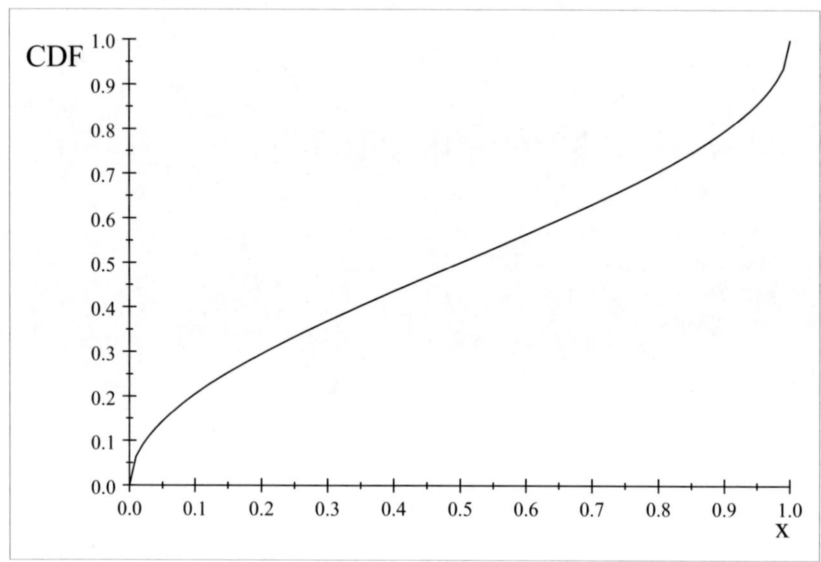

Figure 2.2. cdf – ARS(0,1).

Mean $\frac{b+a}{2}$

Variance $\frac{(b-a)^2}{8}$

First quartile $q_1 = a + (b-a)sin^2\left(\frac{\pi}{8}\right)$

Second quartile $q_2 = a + (b-a)sin^2\left(\frac{\pi}{4}\right)$

Third quartile $q_3 = \text{a+(b-a)}\ sin^2\left(\frac{3\pi}{8}\right)$

2.2. Basic Properties

The characteristic function $\varphi(t)$ is

$$\varphi(t) = \int_a^b \frac{1}{\pi} \frac{e^{itx}}{\sqrt{(x-a)(b-x)}}\, dx$$

Let x= (b-a) z+a, then

$$\varphi(t) = e^{iat}\int_0^1 \frac{1}{\pi}\frac{e^{it(\mathrm{b-a})z}}{\sqrt{(z(1-z))}}\,\mathrm{dz}$$

Let z - $sin^2\theta/2$,

$$= e^{iat}\int_0^{\pi} \frac{1}{\pi}\frac{e^{it(\mathrm{b-a})sin^2\frac{\theta}{2}s}\ \ sin\frac{\theta}{2}\,cos\frac{\theta}{2}}{sin\frac{\theta}{2}cos\frac{\theta}{2}}\,\mathrm{d}\theta$$

$$= e^{ita}\int_0^{\pi}\frac{1}{\pi}e^{it\frac{-(b-a)cos\theta}{2}}\,\mathrm{d}\theta$$

$$= e^{ita}\ I_0(\frac{(b-a)t}{2})$$

where I_0 is the modified Bessel function of order zero and of first kind.

Moment generating function M(t) is

$$\mathrm{M(t)} = \int_a^b \frac{1}{\pi}\frac{e^{tx}}{\sqrt{(x-a)(b-x)}}\,\mathrm{dx}$$

Proceeding as in characteristic function, we obtain

$$\mathrm{M(t)} = e^{ta}\int_0^{\pi}\frac{1}{\pi}e^{\frac{-t(b-a)cos\theta}{2}}$$

$$= e^{ta}I_0(\frac{-i(b-a)t}{2}),$$

Let μ_r = E(X^r), we have In particular

$$\mu_n = \frac{1}{\pi}\int_0^{\pi}(a + (b-a)\,sin^2\frac{\theta}{2})^n d\theta$$

$$\sum_{k=0}^{n} C_k^n a^k\,(b-a)^{n-k}\prod_{j=0}^{n-k-1}\frac{2j+1}{2j+2}$$

$$\mu_1 = \frac{b+a}{2},$$

$$\mu_2 = \frac{3}{8}a^2 + \frac{1}{4}\,\mathrm{ab} + \frac{3}{8}ab^2$$

$$\mu_3 = \frac{5}{16}a^3 + \frac{3}{16}a^2b + \frac{3}{16}ab^2 + \frac{5}{16}b^3$$

$$\mu_4 = \frac{35}{128}a^4 + \frac{5}{32}a^3b + \frac{9}{64}a^2b^2 + \frac{5}{32}ab^3 + \frac{35}{128}b^4$$

If a-0, then

$$\mu_n = \frac{1}{\pi}\int_0^{\pi}(b\,sin^2\frac{\theta}{2})^n d\theta$$

$$=b^n\prod_{k=0}^{n-1}\frac{2k+1}{2k+2}$$

Central moments

$\mu_{c,2n+1}$= E$(X-\frac{a+b}{2})^{2n+1}=0\ for\ n=0,1,2,...$ because X is symmetric with respect to $\frac{a+b}{2}$.

$$\mu_{c,2n}=\text{E}(X-\frac{a+b}{2})^{2n}=\frac{1}{\pi}\int_a^b(x-\frac{a+b}{2})^{2n}\,(x-a)^{-1/2}\,(b-x)^{-1/2}\text{dx}$$

$$=\frac{2}{\pi}\int_{\frac{a+b}{2}}^{b}(x-\frac{a+b}{2})^{2n}\,(x-a)^{-1/2}\,(b-x)^{-1/2}\text{dx}$$

$$=\frac{2}{\pi}\int_0^{\frac{b-a}{2}}t^{2n}((\frac{b-a}{2}+t)^{-1/2}(\frac{b-a}{2}-t)^{-1/2}\,\text{dt}$$

$$=\frac{2}{\pi}\int_0^{\frac{b-a}{2}}t^{2n}((\frac{b-a}{2})^2\text{-}t^2)\text{dt}$$

$$=\frac{2}{\pi}((\frac{b-a}{2})^{2n}\int_0^1 u^{2n}(1-u^2)^{-1/2}d\text{u}$$

$$=\frac{1}{\pi}\frac{(b-a)^{2n}}{4^n}\int_0^1 w^{n-\frac{1}{2}}(1-w)\text{dw}$$

$$=\frac{1}{\pi}\frac{(b-a)^{2n}}{4^n}\,\text{B}\,(\text{n}+\frac{1}{2},\frac{1}{2})$$

$$=\frac{1}{\sqrt{\pi}}\frac{(b-a)^{2n}}{4^n}\frac{\Gamma(n+\frac{1}{2})}{\Gamma(n+1)}.$$

2.3. Distributional Properties

If X has ARS (a,b), then c X+d is ARS (ac+d.bc+d)

If X is Cauchy (0,1), then $\frac{!}{1+X^2}$ is ARS (0,1).

If U has the standard uniform distribution, then X = $sin^2\left(\frac{\pi U}{2)}\right)$ is ARS (0,1)

2.4. Random Number Generation

Generate a random number U from a uniform distribution U(0,1).

Set X = a +(b-a)(sin $(\frac{\pi U}{2}))^2$, $-\infty<a<b<\infty$, then X is an ARS (a,b) random variable.

Chapter 3

Bernoulli Distribution (BER(p))

3.1. Introduction

PMF $\quad p(x) = p$ if $x = 1$
$\quad = 1-p$ if $x=0$, $0<p<1$.

The PMFs for p=0.3 and p=0.6 are given in figures 3.1.1. and 3.1.2

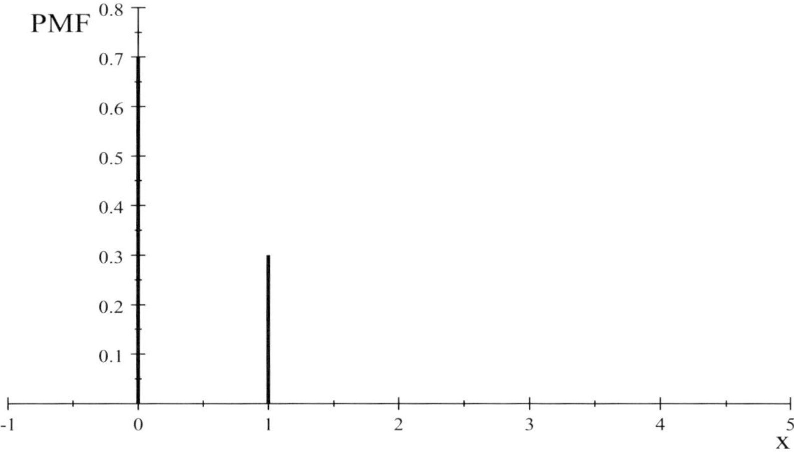

Figure 3.1.1. PMF-BER(0.3).

The cumulative distribution function (CDF) P(x) is given by

$P(x) = 0$, if $x<0$
$= 1-p$ if $0\leq x < 1$
$=1$ if $x\geq 1$

The cdfs for p=04 and p=0.6 are given in figures 3.1.3, and 3.1.4.

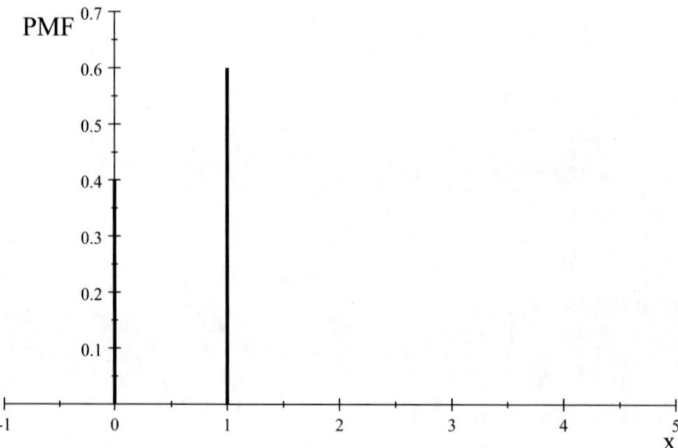

Figure 3.1.2. PMF-BER(0.6).

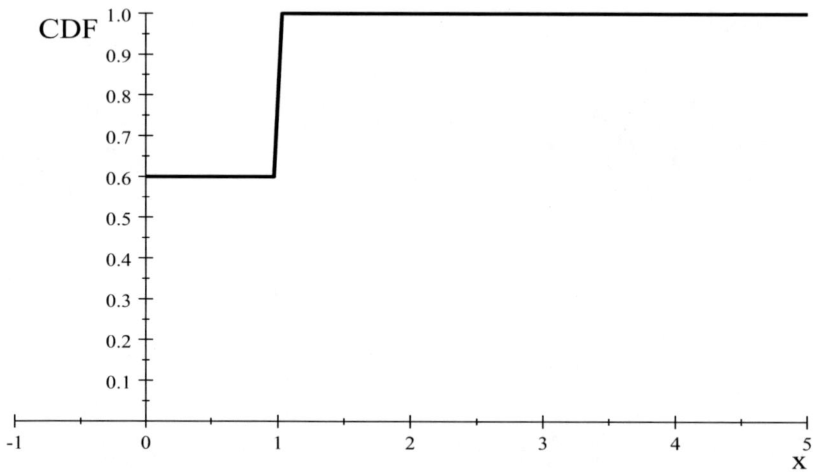

Figure 3.1.3. CDF- p=0.4.

Mean, $\mu = \text{p}$

Variance . α^2 = pq

n-th moment E(X^n) is

E(X^n) = $(0)^n$ (1-p) +$(1)^n p$ = p

Coefficient of skewness $\frac{1-2p}{\sqrt{p(1-p)}}$

Kurtosis $\frac{1}{p(1-p)}$

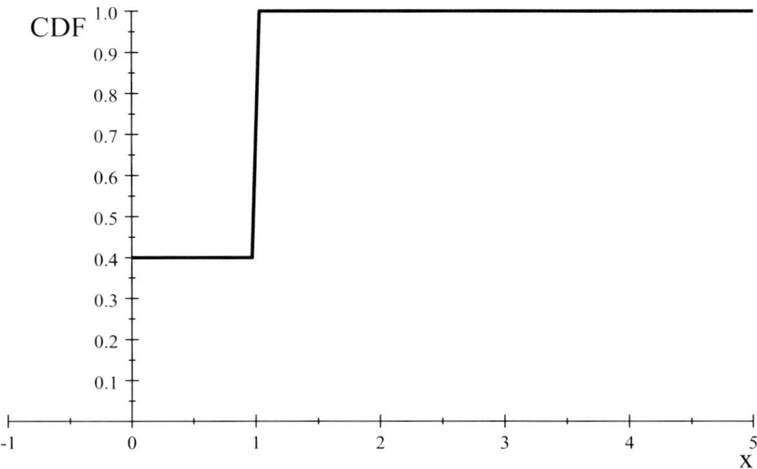

Figure 3.1.4. CDF-BER(0.6).

3.2. Basic Properties

The moment generating function M(t) is

$M(t) = E(e^{tx}) = q + e^{t}, q = 1-p.$

Central moments

$\mu_n = E(X-p)^n = (-p)^n(1-p) + (1-p)^n p$
$\mu_1 = -(p)(1-p) + p(1-p) = 0$
$\mu_2 = p^2(1-p) + (1-p)^2 p = p(1-p)$
$\mu_3 = -p^3(1-p) + (1-p)^3 p = p(1-p)(1-2p)$

The characteristic function $\phi(t)$ is

$\phi(t) = q + e^{it}$

Probability generating function.

$q + ps$

3.3. Distributional Properties

The distribution of the sum of independent Bernoulli random is Binomial distribution.

If $X_1, X_2, \dots, X_n$ are independent BER(p) and
$M_n = \max(X_1, X_2, \dots, X_n)$, then
$M_n = \text{BER}(R_n)$,
$where\ R_n = 1 - (1 - p)^n$

3.4. Random Number Generation

Generate a random number from a uniform, UN (0,1) distribution. Select a p, 0<p<1, if the random number is less than p, then the Bernoulli of random variables that are less than p, the Bernoulli random is 1, otherwise it is 0.

Chapter 4

Beta Distribution (BE (m,n))

4.1. Introduction

Pdf $\frac{1}{B(m,n)} x^{m-1} (1-x)^{n-1}$

$0 \leq x \leq 1, m \geq 0, n \geq 0.$

The pdfs of BE (2,3), BE (3,7) and BE (4,10) are given in the figure 4.1.1.

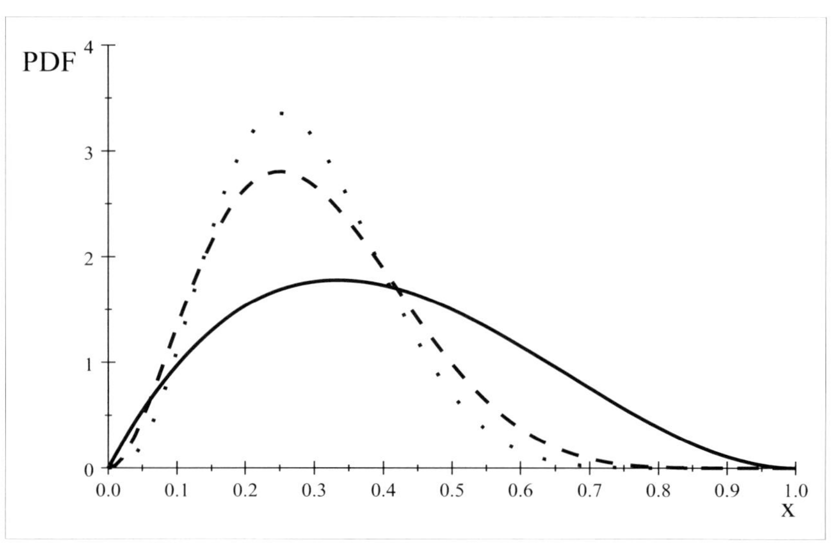

Figure 4.1.1. BE (2,3)-Solid, BE (3,7)-dash and BE (4,10)-dots.

Cdf $\frac{B(m,n,x)}{B(m,n)}$

$\mathrm{B(m,n,x)} = \int_0^x u^{m-1} (1-u)^{n-1}\, \mathrm{du}$

The cdfs of BE (2,3), BE (3,7) and BE (4,10) are given in Figure 4.1.2.

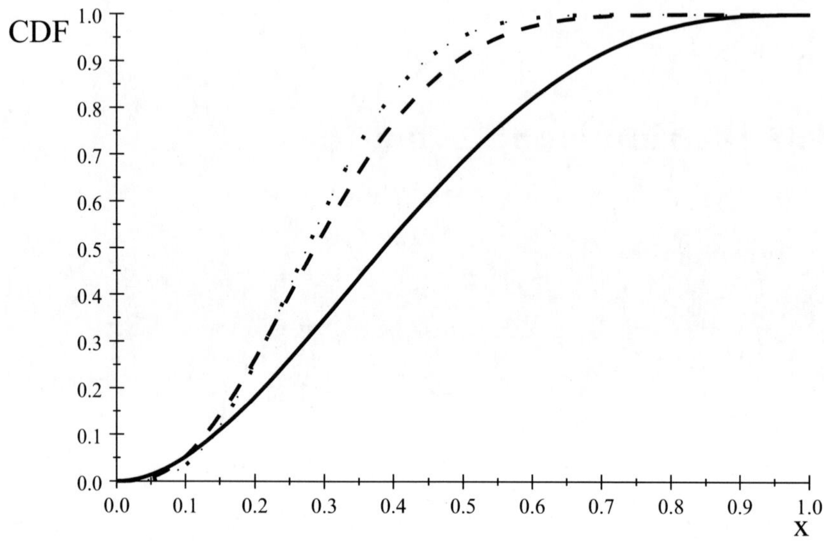

Figure 4.1.2. cdfs BE (2,3)-solid, BE (3,7)-dash and BE (4,10) -dots.

Mean	$\frac{m}{m+n}$
Variance	$\frac{mn}{(m+n)^2(m+n+1)}$
Coefficient of variation	$(\frac{n}{m(n+m+1)})^{1/2}$

4.2. Basic Properties

Moment about the origin $E(X^k)$

$$E(X^k) = \frac{B(m+k,n)}{B(m,n)}, k = 1,2,...$$

$$E(X^2) = \frac{m(m+1)}{(m+n)(m+n+1)}$$

$$E(X^3) = \frac{m(m+1)(m+2)}{(m+n)(m+n+1)(m+n+2)}$$

$$E(\ln x) = \int_0^1 \frac{d}{dm} \frac{x^{m-1}(1-x)^{n-1}}{B(m,n)}.$$

$$= \frac{1}{B(m,n)} \frac{d}{dm} \int_0^1 x^{m-1}(1-x)^{n-1}$$

$=\frac{1}{B(m,n)}\frac{d}{dm}B\,(m,n)$
$=\frac{d}{dm}\,ln\Gamma(m)-\frac{d}{dm}\,ln\Gamma(m+n)$
$=\Psi(m)-\Psi(m+n)$,

where Ψ is Digamma function

4.3. Distributional Properties

Moment generating function, $M_{m,n}\,(t)$

$$M_{m,n}\,(t)=\sum_{k=0}^{\infty}\frac{B(m+k,n)}{B(m,n)}\frac{t^k}{k!}$$

Characteristic function, ϕ(m.n,t)

$$\phi(m.n,t)=\sum_{k=0}^{\infty}\frac{B(m+k,n)}{B(m,n)}\frac{(it)^k}{k!}$$
B(m, n) = B(n, m)
B(1/2.1/2) = π

If m=1/2. n=1/2, then BE (1/2,1/2) is distributed as arcsine distribution.

If m=1, n=1, then BE (1,1) is distributed as uniform distribution, UN (0,1).

If m=1, then, BE(1,n) is distributed as power function distribution, POW (0,1, n).

If X is distributed as BE (m, n), then 1-X is distributed as BE (n,m).

The r-th order statistics from Uniform distribution, UN (0,1) is distributed as

BE (r,n-r+1).

If X is distributed as Beta, BE (m, n), then $\frac{nX}{m(1-X)}$ is distributed as F(2m,2n).

If X is distributed as uniform, U (0,1), then X^2 is distributed as a BE (1/2,1).

4.4. Random Number Generation

Take two integers m and n

Generate $U_1, U_2, \ldots, U_m$ uniform, UN (0,1) random variables.
Generate $V_1, V_2, \ldots V_n$ uniform, UN (0,1) random variables.

Set $X = \Pi_{i=1}^{m} U_i$

Set $X = \Pi_{i=1}^{n} V_i$

Set $W = \frac{X}{X+Y}$

Then W is BE (m, n).

Chapter 5

Binomial Distribution (BN (n,p))

5.1. Introduction

PMF $\quad P(X{=}k){=}\binom{n}{k}p^k(1-p)^{n-k}$, k=0,1,2,…,n.

The PMFs of BN (5,0.3) and BN (5,0.5) are given in figures 5.1.1 and 5.1.2.

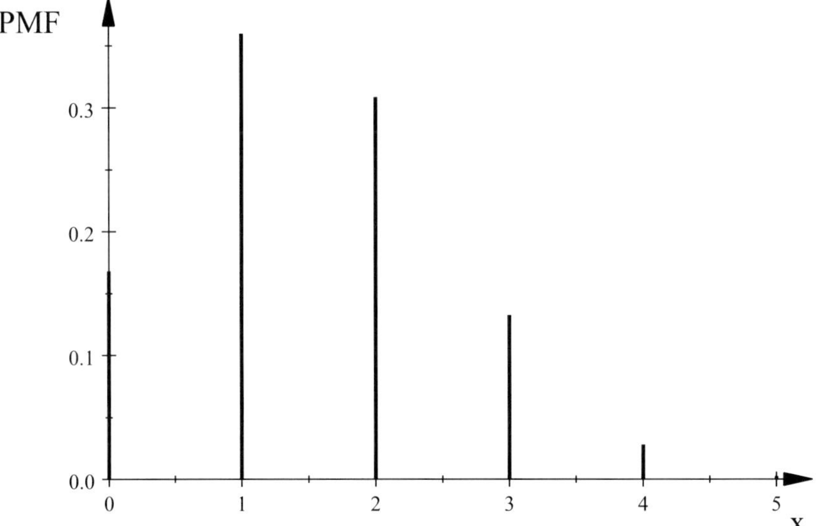

Figure 5.1.1. PMF – BN (5,0.3).

Cdf $\quad P(X\leq x) = \sum_{k=o}^{x}\binom{n}{k}\, p^k(1-p)^{n=k}$

The cdfs of BN (5,0.3) and BN (5,0.5) are given in figures 5.1.3 and 5.1.4.

$\sum$ 1 p

x k 0 $p^x($ − n x

= n k)) −

(

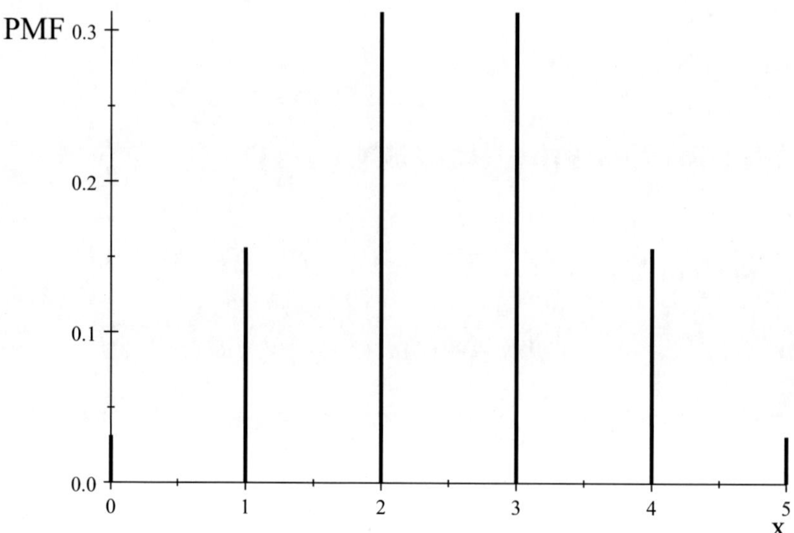

Figure 5.1.2. PMF- BN (5,0.5).

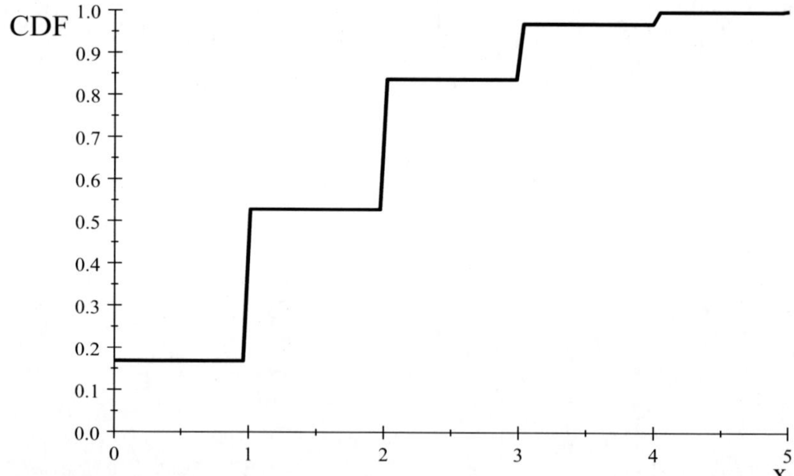

Figure 5.1.3. Cdf- BN (5,0.3).

Mean	np
Variance	np(1-p)
Coefficient of variation	$\sqrt{(\frac{1-p}{np})}$

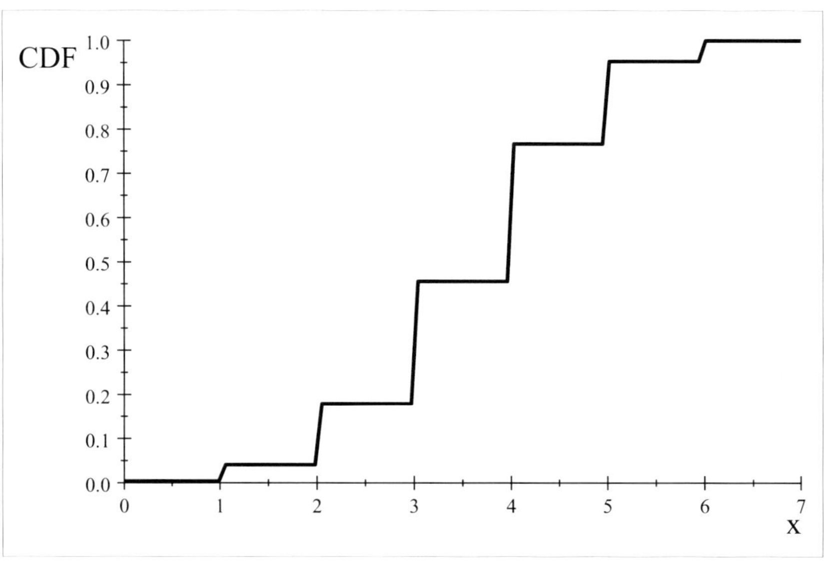

Figure 5.1.4. Cdf- BN (6,0.6).

5.2. Basic Properties

$$\frac{P(X-x+1)}{P(X=x)} = \frac{(n-x)p}{(x+1)q}$$

Moment generating function $(q + pe^t)^n$

$E(X^2) = np(1-p) + n^2p^2$

In general

$E(X^m)=\sum_{=0}^{m}\{{m \atop k}\}m^{(k)}p^k$

where $\{{m \atop k}\}$ *is the Stirling Nmbers of the second kind ,*

$$\{{m \atop k}\} = \sum_{i=o}^{k} \frac{(-1)^{k-i} i^m}{(k-i)!\, i!}$$

$m^{(k)}$=m(m-1)....(m-k+1) with $m^{(0)}$=0.

Factorial moment	E(X(X-1)…(X-k+1) =n(n-1)…(n-k+1)p^k, k=1,2,…
Central moment	
μ_r	$E(X-np)^r$
μ_2	np(1-p)
μ_3	np(1-p)(1-2p)
μ_4	$(np(1-p))^2$+np(1-p) (1-6p(1-p))
μ_{r+1}	$np(1-p)^2(nr\,\mu_{r-1}+\frac{d\mu_r}{dp})$

5.3. Distributional Properties

Probability generating function	$q+pt)^n$
$Characteristic\ function$	$(q+pe^{it})^n$

If X is distributed as BN (1,p), then X is known as Bernoulli random variable, BER (p).

If X is distributed as BN (n,p). Then $\lim_{n\to\infty}\frac{X-np}{\sqrt{(npq)}}$ converges to N (0,1) in the distribution.

5.4. Random Number Generation

Generate n random number for uniform, U (0,1) distribution.

Select a number p, 0<p<1.

The number of random Numbers less than p, is BN (n,p).

Chapter 6

Birnbaum-Saunders Distribution (BS($\alpha . \beta, \lambda$))

6.1. Introduction

Birnbaum-Saunders Distribution or BS (3P) Distribution

$$pdf\ f_X(x;\ \alpha,\ \beta,\ \lambda) = \frac{1}{2\alpha(x-\lambda)}\left[\sqrt{\frac{x-\lambda}{\beta}} + \sqrt{\frac{\beta}{x-\lambda}}\right] \times$$

$$\phi\left[\frac{1}{\alpha}\left\{\sqrt{\frac{x-\lambda}{\beta}} - \sqrt{\frac{\beta}{x-\lambda}}\right\}\right], \tag{6.1}$$

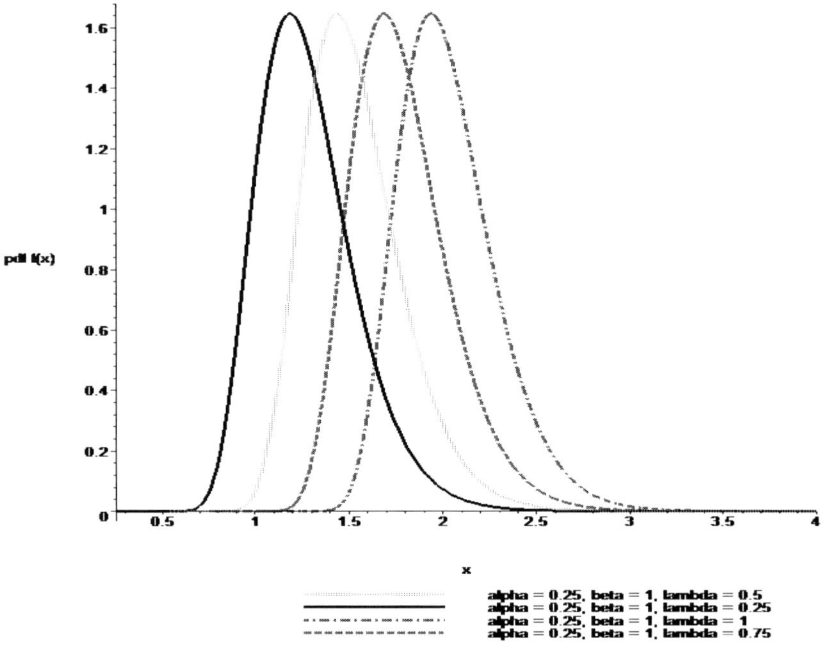

Figure 6.1. (a). PDF of $\boldsymbol{X} \sim \mathrm{BS}(\alpha, \beta, \lambda)$.

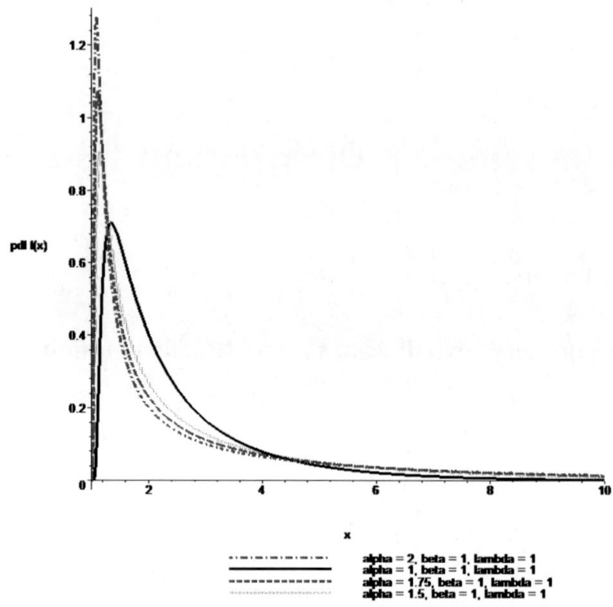

Figure 6.1. (b). PDF of $\boldsymbol{X} - \mathrm{BS}(\alpha, \beta, \lambda)$.

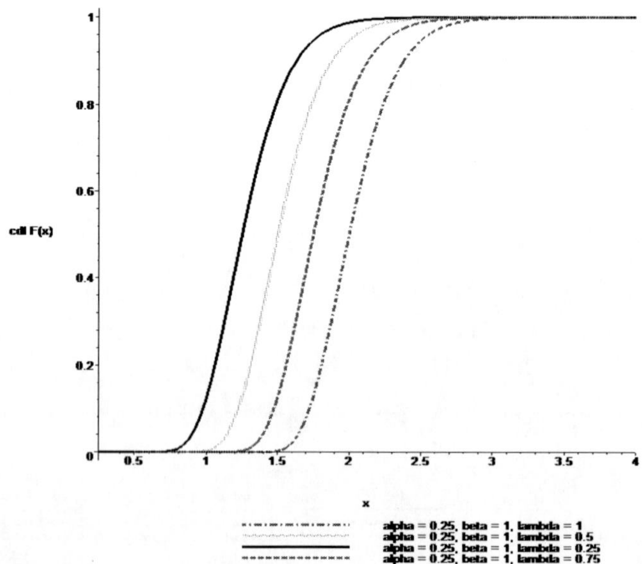

Figure 6.1. (c). PDF of $\boldsymbol{X} - \mathrm{BS}(\alpha, \beta, \lambda)$.

$$cdf\ F_X(x;\ \alpha,\ \beta,\ \lambda) = \Phi\left[\frac{1}{\alpha}\left\{\sqrt{\frac{x-\lambda}{\beta}} - \sqrt{\frac{\beta}{x-\lambda}}\right\}\right], \tag{6.2}$$

where $x > \lambda \geq 0$, $\alpha > 0$, $\beta > 0$ are Location, shape and scale parameters, respectively, $\phi()$ is the probability density function (pdf) and $\Phi()$ is the cumulative distribution function (cdf) of the normal distribution. The graphs of the pdf and cdf of the Birnbaum-Saunders distribution, $X \sim BS(\alpha,\ \beta,\ \lambda)$, are provided for different values of α, β and λ in figures 6.1 (a, b) and 6.2 (a, b), respectively.

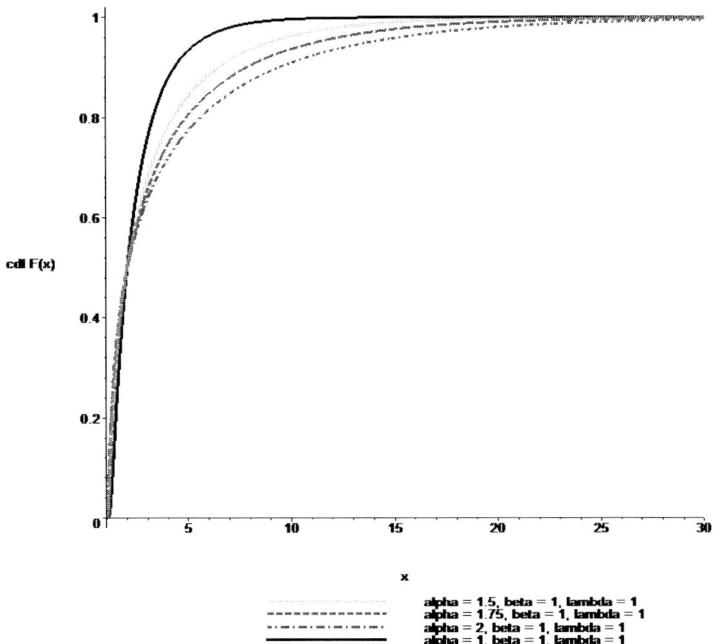

Figure 6.2. (a). CDF of $\boldsymbol{X} - \mathrm{BS}(\alpha,\beta,\lambda)$.

The hazard function (hf) of the Birnbaum-Saunders distribution, $X \sim BS(\alpha,\ \beta,\ \lambda)$, is given by

$$hf\ h_X(x;\alpha,\beta,\lambda) = \frac{\frac{1}{2\alpha(x-\lambda)}\left[\sqrt{\frac{x-\lambda}{\beta}} + \sqrt{\frac{\beta}{x-\lambda}}\right] \times \phi\left[\frac{1}{\alpha}\left\{\sqrt{\frac{x-\lambda}{\beta}} - \sqrt{\frac{\beta}{x-\lambda}}\right\}\right]}{1 - \Phi\left[\frac{1}{\alpha}\left\{\sqrt{\frac{x-\lambda}{\beta}} - \sqrt{\frac{\beta}{x-\lambda}}\right\}\right]}. \tag{6.3}$$

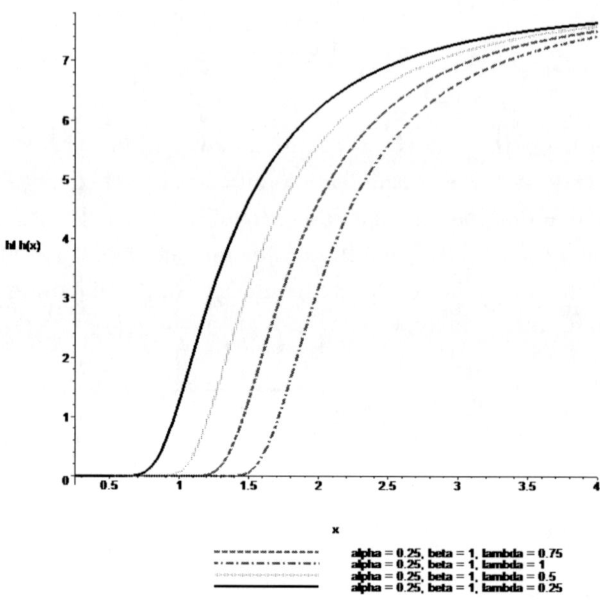

Figure 6.3. (b). hf of $\boldsymbol{X} \sim \mathrm{BS}(\alpha, \beta, \lambda)$.

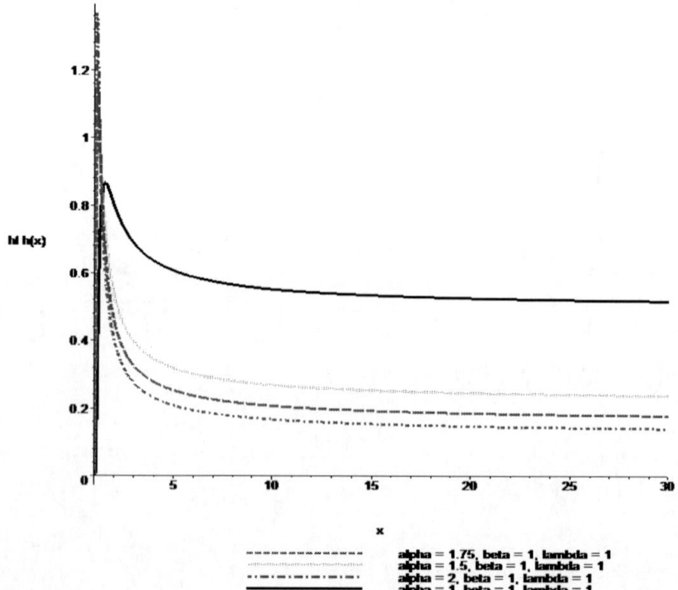

Figure 6.3. (b). hf of $\boldsymbol{X} \sim \mathrm{BS}(\alpha, \beta, \lambda)$.

The graphs of the hazard function (hf) (6.3) of $X \sim BS(\alpha,\ \beta,\ \lambda)$are provided for different values of α, β and λ in figures 6.3 (a, b).

6.2. Basic Properties

The mean of $X\ \sim\ BS\big(\alpha, \beta, \lambda\big)$, $\mathbb{E}(X) = \mu$, is defined by

$$\mu\ =\ \frac{\beta}{2}(\alpha^2\ +\ 2)\ +\ \lambda. \tag{6.4}$$

The variance, $\mathbb{E}(X^2) - (\mathbb{E}(X))^2 = \sigma^2$, is defined by

$$\sigma^2 = (\alpha\beta)^2 \left(1\ +\ \frac{5\alpha^2}{4}\right). \tag{6.5}$$

The skewness, $\alpha_3\ = \frac{\mathbb{E}(X-\mu)^3}{\sigma^3}$, is defined by

$$\alpha_3 = \frac{4\alpha\left(11\alpha^2\ +\ 6\right)}{(5\alpha^2\ +\ 4)^{3/2}} \tag{6.6}$$

The kurtosis, $\alpha_4\ = \frac{\mathbb{E}(X-\mu)^4}{\sigma^4}$, is given by

$$\alpha_4 = \ = 3 + \frac{6\alpha^2\left(93\alpha^2 + 40\right)}{\left(5\alpha^2 + 4\right)^2}\,. \tag{6.7}$$

(i) When $\lambda = 0$, the Birnbaum-Saunders distribution, $X, BS(\alpha,\ \beta,\ \lambda)$, reduces to the BS (2P) distribution.

(ii) When $\lambda = 0, \beta = 1$, the Birnbaum-Saunders distribution, $X \sim BS(\alpha,\ \beta,\ \lambda)$, is called the standard Birnbaum-Saunders or the standard fatigue life distribution.

Birnbaum-Saunders (2P) Distribution

$$pdf\ f_X(x;\alpha,\beta) = \frac{1}{2\sqrt{2\pi}\alpha\beta}\left[\left(\frac{\beta}{x}\right)^{\frac{1}{2}} + \left(\frac{\beta}{x}\right)^{\frac{3}{2}}\right] exp\left[-\frac{1}{2\alpha^2}\left(\frac{x}{\beta} + \frac{\beta}{x} - 2\right)\right], \tag{6.8}$$

and

$$cdf\ F_X(x;\alpha,\beta) = \Phi\left(\frac{1}{\alpha}\left[\left(\frac{x}{\beta}\right)^{\frac{1}{2}} - \left(\frac{\beta}{x}\right)^{\frac{1}{2}}\right]\right), \tag{6.9}$$

where $0 < x < \infty$, $\alpha, \beta > 0$, and $\Phi(.)$ is the cumulative distribution function (cdf) of the normal distribution.

6.3. Distributional Properties

The mean of $X \sim BS(\alpha, \beta)$, $\mathbb{E}(X) = \mu$, is defined by

$$\mu = \beta\left(1 + \frac{\alpha^2}{2}\right). \tag{6.10}$$

The variance, $\mathbb{E}(X^2) - (\mathbb{E}(X))^2 = \sigma^2$, is defined by

$$\sigma^2 = (\alpha\beta)^2\left(1 + \frac{5\alpha^2}{4}\right). \tag{6.11}$$

The skewness, $\alpha_3 = \frac{\mathbb{E}(X-\mu)^3}{\sigma^3}$, is defined by

$$\alpha_3 = \frac{4\alpha(11\alpha^2+6)}{(5\alpha^2+4)^{\frac{3}{2}}}. \tag{6.12}$$

The kurtosis, $\alpha_4 = \frac{\mathbb{E}(X-\mu)^4}{\sigma^4}$, is given by

$$\alpha_4 = 3 + \frac{6\alpha^2(93\alpha^2+40)}{(5\alpha^2+4)^2}. \tag{6.13}$$

Remark 1.

1. If $X \sim BS(\alpha, \beta)$, then $X^{-1} \sim BS(\alpha, \beta^{-1})$.
2. If $X \sim BS(\alpha, \beta)$, then $Z = \frac{1}{\alpha}\left(\sqrt{\frac{X}{\beta}} - \sqrt{\frac{\beta}{X}}\right) \sim N(0,1)$.
3. Let $X \sim BS(\alpha, \beta)$ and $T = \frac{1}{\alpha}\left[\left(\frac{X}{\beta}\right)^{\frac{1}{2}} - \left(\frac{X}{\beta}\right)^{-\frac{1}{2}}\right]$, or, equivalently,

$X = \beta\left(1 + \alpha T^2 + \alpha T(1 + T^2)^{\frac{1}{2}}\right)$. Then T is distributed normally with a mean of zero and a variance of $\frac{\alpha^2}{4}$.

6.4. Random Number Generation

Consider a random variable, U, drawn from the uniform distribution in the interval $(0, 1)$. Then the random variable X defined by

$$X = \left[\frac{\alpha}{2}F^{-1}(U) + \sqrt{\left(\frac{\alpha}{2}F^{-1}(U)\right)^2 + 1}\right]^2 \beta$$

has a BS (2P), $X \sim BS(\alpha, \beta)$, distribution with the parameters $\alpha > 0$ and $\beta > 0$.

Chapter 7

Burr (3P) Distribution

7.1. Introduction

$$pdf \quad f(x) = \frac{\alpha k\left(\frac{x}{\beta}\right)^{\alpha - 1}}{\beta\left(1+\left(\frac{x}{\beta}\right)^{\alpha}\right)^{k + 1}}, \tag{7.1}$$

where $k(> 0)$: shape parameter; $\alpha(> 0)$: shape parameter; $\beta(> 0)$: scale parameter; and domain: $0 \leq x < \infty$.

The pdfs of Burr (3P) pdf (7.1) are given in figure 7.1.

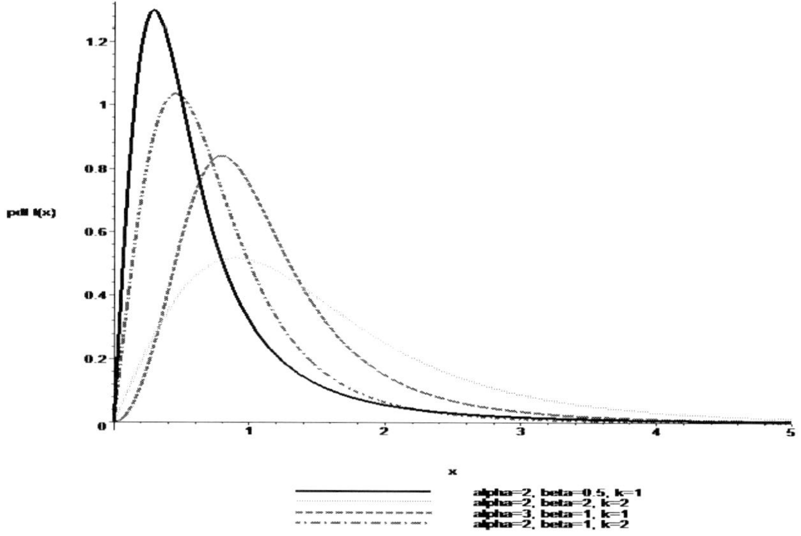

Figure 7.1. PDF- Burr (3P).

$$cdf\ F(x) = 1 - \left(1 + \left(\frac{x}{\beta}\right)^{\alpha}\right)^{-k}. \tag{7.2}$$

The pdfs of Burr (3P) cdf (7.2) are given in figure 7.2.

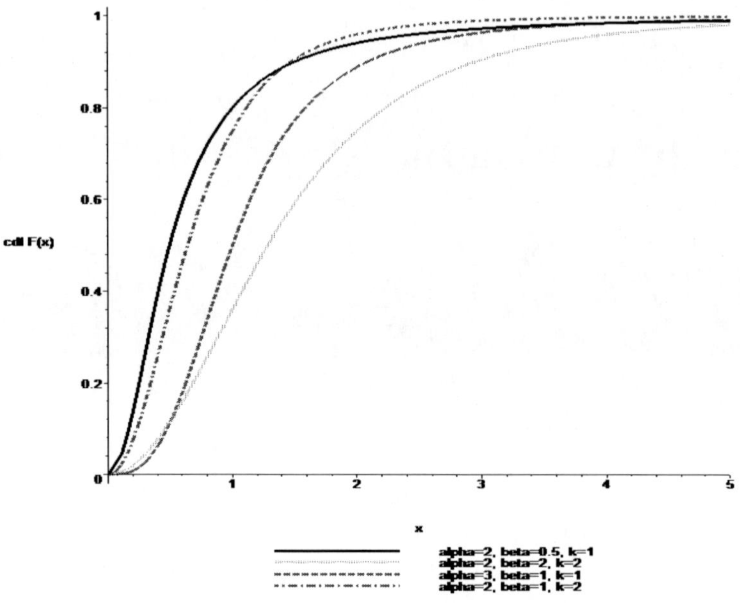

Figure 7.2. CDF Burr (3P).

7.2. Basic Properties

jth moment of Burr (3P) = $E(X^j) = (k\beta^j)\, B\left(1 + \frac{j}{\alpha},\, k - \frac{j}{\alpha}\right), j < \alpha,$ $j < \beta$

where $B(.)$ denotes the complete beta function.

Mean (or, First Moment) of Burr (3P) = $E(X)$

$$= (k\beta)B\left(1 + \frac{1}{\alpha},\, k - \frac{1}{\alpha}\right),\ 0 < \frac{1}{\alpha} < k.$$

Second Moment of Burr (3P) = $E(X^2) = (k\beta^2)B\left(1 + \frac{2}{\alpha},\, k - \frac{2}{\alpha}\right).$

jth (Central) Moment of Burr (3P) = $E[X - E(X)]^j$

$$= \sum_{m=0}^{j}(-1)^m \binom{j}{m}\left((k\beta)B\left(1 + \frac{1}{\alpha},\, k - \frac{1}{\alpha}\right)\right)^m,$$

where $j - m < \alpha, j - m < \beta.$

$$\text{Skewness} = \frac{\sum_{m=0}^{3}(-1)^m\binom{3}{m}\left((k\beta)B\left(1+\frac{1}{\alpha}, k-\frac{1}{\alpha}\right)\right)^m (k\beta^{3-m})\, B\left(1+\frac{3-m}{\alpha}, k-\frac{3-m}{\alpha}\right)}{\left((k\beta^2)\left\{B\left(1+\frac{2}{\alpha}, k-\frac{2}{\alpha}\right)-(k)\left[B\left(1+\frac{1}{\alpha}, k-\frac{1}{\alpha}\right)\right]^2\right\}\right)^{\frac{3}{2}}},$$

$$\text{Kurtosis} = \frac{\sum_{m=0}^{4}(-1)^m\binom{4}{m}\left((k\beta)B\left(1+\frac{1}{\alpha}, k-\frac{1}{\alpha}\right)\right)^m (k\beta^{4-m})\, B\left(1+\frac{4-m}{\alpha}, k-\frac{4-m}{\alpha}\right)}{\left((k\beta^2)\left\{B\left(1+\frac{2}{\alpha}, k-\frac{2}{\alpha}\right)-(k)\left[B\left(1+\frac{1}{\alpha}, k-\frac{1}{\alpha}\right)\right]^2\right\}\right)^{2}}.$$

Mode of Burr (3P) $= x = (\beta)\left(\frac{\alpha-1}{\alpha k+1}\right)^{\frac{1}{\alpha}}, \alpha k \geq 1, k > 0, \alpha > 0, \beta > 0.$

Median of Burr (3P) $= x = (\beta)\left((2)^{\frac{1}{k}} - 1\right)^{\frac{1}{\alpha}}.$

Quantile Function of the Burr (3P) Distribution $= Q(x;\ k,\alpha,\beta) = \beta\left((1-u)^{-\frac{1}{k}} - 1\right)^{\frac{1}{\alpha}},$

where $u \in (0,1)$

Moment Generating Function (MGF) of Burr (3P) $= M_X(t) = E(e^{tX})$

$$= \sum_{j=0}^{\infty}\frac{(t)^j}{j!}E(X^j) = \sum_{j=0}^{\infty}\frac{(t)^j}{j!}(k\beta^j)\, B\left(1+\frac{j}{\alpha}, k-\frac{j}{\alpha}\right), t \in \mathbb{R}.$$

Note: We can obtain all moments of X^j from its MGF as follows:

$$E(X^j) = \frac{d^j}{dt^j}M_X(t)\Big|_{t=0}.$$

Characteristic Function of Burr (3P) $= \phi_X(t) = M_X(it) = E(e^{itX})$ $= \sum_{j=0}^{\infty}\frac{(it)^j}{j!}E(X^j) = \sum_{j=0}^{\infty}\frac{(it)^j}{j!}(k\beta^j)\, B\left(1+\frac{j}{\alpha}, k-\frac{j}{\alpha}\right), t \in \mathbb{R},$ where $i = \sqrt{-1}$ is the imaginary number, $i^2 = -1$. rth Cumulant of Burr (3P) $= \kappa_r = \frac{1}{i^r}\left[\frac{d^r(ln(\phi_X(t)))}{dt^r}\right]_{t=0}$, $r = 1, 2, \dots,$ from which, by successive differentiation, it can be easily seen that $\kappa_1 = E(X)$, $\kappa_2 = Var(X)$, $\kappa_3 = E[X - E(X)]^3$, etc.

7.3. Distributional Properties

The Burr (3P) distribution is related to various other well-known distributions, depending on the values of its parameters.

i. *Pareto Distribution:* If $\alpha = 1$, the Burr (3P) distribution simplifies to a *Pareto distribution* with two parameters, which is also known as the *Lomax distribution.*
ii. *Loglogistic Distribution:* For $k = 1$, the Burr (3P) distribution becomes *the Loglogistic distribution,* which is also known as the *Fisk distribution.*
iii. *Dagum (3P) Distribution:* If X has the Burr (3P) distribution, then the Dagum (3P) distribution is the distribution of $\frac{1}{X}$, which is also known as the *inverse Burr (3P) distribution.*
iv. *Weibull Distribution:* With certain parameter choices, the Burr (3P) distribution can approximate the *Weibull distribution,* particularly when modeling heavy-tailed or skewed data.
v. *Generalized Pareto Distribution (GPD):* The Burr (3P) distribution can also be seen as a generalization of the Pareto distribution and is connected to the Generalized Pareto Distribution (GPD).

7.4. Random Number Generation

Random numbers from a Burr (3P) distribution can be generated using the inverse transform method, which utilizes the quantile function derived above.

Steps to Generate Random Numbers

1. Generate a random uniform number u from $U(0, 1)$.
2. Apply the quantile function $Q(x;\ k, \alpha, \beta)$ to get a sample x from the Burr (3P) distribution:

$$x = \beta\left((1 - u)^{-\frac{1}{k}} - 1\right)^{\frac{1}{\alpha}}.$$

This method ensures that the generated random numbers follow the desired Burr (3P) distribution.

Chapter 8

Cauchy Distribution (CA(μ,σ))

8.1. Introduction

Pdf $\quad \frac{1}{\pi\sigma(1+(\frac{x-\mu}{\sigma})^2)} \quad -\infty < \mu < x < \infty. \sigma > 0.$

The pdfs of CA (0,1/2) CA (0,1) and CA (0,3) are given in figure 8.1.

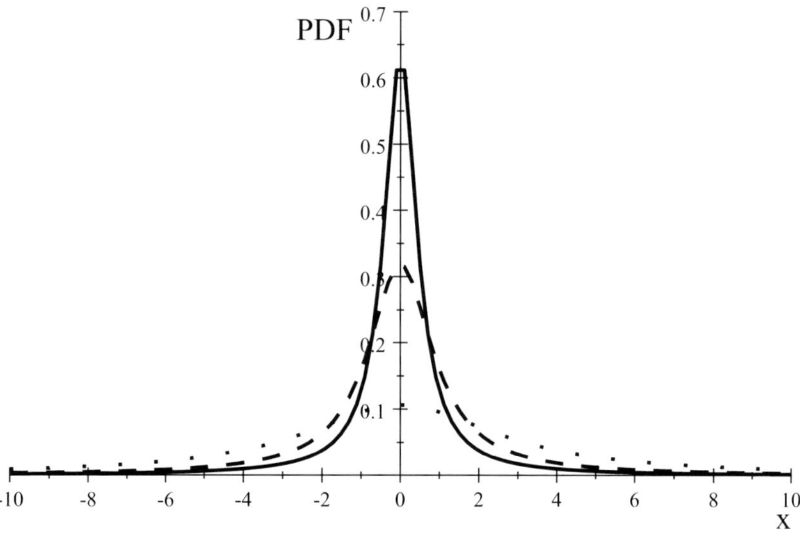

Figure 8.1. PDFs – CA (0,1/2)-solid, C A (0,1)-dash and CA (0,3)-dots.

CDF $\quad \frac{1}{2}+\frac{1}{\pi}\arctan\left(\frac{x-\mu}{\sigma}\right)$

The CDF of CA (0.1/2), CA (0,1) and CA (0,3)-are given in figure 8.2.

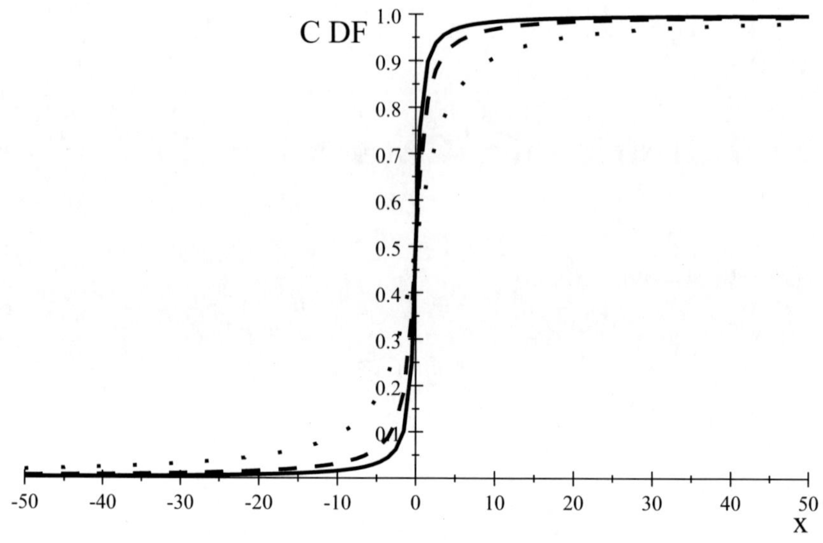

Figure 8.2. CDFs- CA (0.1/2)-solid, CA (0,1)=dash and CA (0,3)-dots.

Mean	Does not exist
Median	μ
Mode	μ
First quartile	$\mu - \sigma$
Third quartile	$\mu + \sigma$
Point of inflection	$\mu \pm \dfrac{\sigma}{\sqrt{3}}$

8.2. Basic Properties

The percentile points of CA (0,1) are given in Table 8.1.

Table 8.1. Percentage points of CA (0,1),CA (0.2) and CA (),3)

P	CA (0,1)	CA (0,2)	CA(0,3)
0.05	-6.3138	-12.6280	-18.9410
0.1	-3.0777	-6.1554	-9.2331
0.2	-1.3764	-2.7524	-4.1291
0.3	-0.7266	-1,4531	-2.1798
0.4	-0.3248	-0.6498	-0.9748
0.5	0	0	0
0.6	0.3248	0.6438	0.9748
0.7	0.7266	1.4531	2.1798
0.8	1.3764	2.7524	4.1291
0.9	3.0777	6.1554	9.2331
0.95	6.3136	12.6280	18.9410

Mean Does not exist

Moment generating function - does not exist

Characteristic function $e^{i\mu t-\sigma|t|}$

Hazard rate function

$$\frac{1}{\pi\sigma(1+(\frac{x-\mu}{\sigma})^2)(\frac{1}{2}-\frac{1}{\pi}\tan^{-1}(\frac{x-\mu}{\sigma}))}$$

8.3. Distributional Properties

If X and Y are independent N(0,1), then X/Y is CA (0,1).

If X is CA (0,1), then $2X/(1-X^2)$ is CA (0,1).

If $X_1, X_2,\ldots,X_n$ are independent CA (0,1), then

$\sum_{i=1}^{n} a_i X_i$ is

CA (0, $\sum_{i=1}^{n}|a_i|$

If X is CA (μ,σ), then 1/X, is CA ($\frac{\mu}{\mu^2+\sigma^2},\frac{\sigma}{\mu^2+\sigma^2}$)...

If X is CA (a,b), then

$$X \underset{=}{d} Y_1 + Y_2 + \ldots + Y_n,$$

where Y_1, Y_2, .Y_n are independent and identically distributed as CA (a/n. b/n).

8.4. Random Number Generation

Let U be a uniform random variable, UN (0,1).

Set X = Tan(π(U - $\frac{1}{2}$)).

Then X is CA (0,1).

Chapter 9

Chi-Square Distribution (CHI($\mu.\sigma,n$))))

9.1. Introduction

Pdf $$\frac{1}{2^{\frac{n}{2}}\Gamma\left(\frac{n}{2}\right)}e^{-\frac{1}{2}\left(\frac{x-\mu}{\sigma}\right)}\left(\frac{x-\mu}{\sigma}\right)^{\frac{n}{2}-1}, x > \mu, \sigma > 0,$$

The parameter n is known as degrees of freedom.

The pdfs of CHI (0,1,6), CHI (0,1,12), CHI (0,1,16) are given in figure 9.1.1.

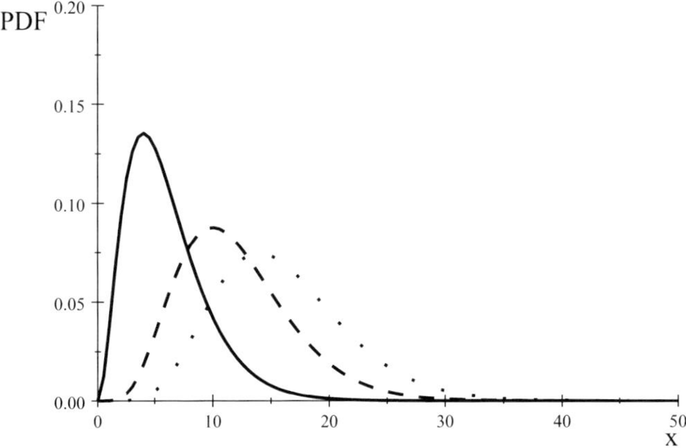

Figure 9.1.1. pdfs - CHI (0,1,6)-solid, CHI (0,1,12)-dash and CHI (0,1,16)-dots.

Cdf of CHI (0,1,n) $$\frac{\Gamma(\frac{r}{2},\frac{x}{2})}{\Gamma(\frac{r}{2})},$$

where

$$\Gamma(m,x) = \int_0^x u^{m-1}\, e^{-u}\, \mathrm{du}$$

where $\Gamma(m,x) = \dfrac{1}{\Gamma(m)} \int_0^x x^{n-1} e^{-x} dx$

The cdfs of CHI (0,1,6), CHI (0,1,12) and CHI (0,1,16) are given in figure 9.1.2.

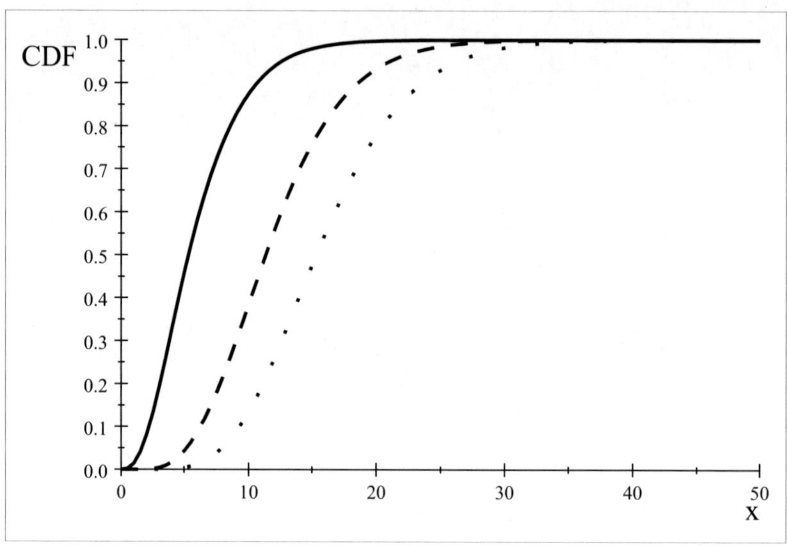

Figure 9.1.2. Cdfs- CHI (0.1.6)-solid. CHI (0,1,2)-dash and CHI (0,1,16)-dots.

Mean	$\mu + n\sigma$
Variance	$2n\sigma^2$
Mode	$\mu + (n-2)\sigma$
Coefficient of variation	$\dfrac{\sigma\sqrt{2n}}{\mu + n\sigma}$

9.2. Basic Properties

The moment generating function $\qquad e^{\mu t}(1-2\sigma t)^{-n/2}, t < \dfrac{1}{2\sigma}$

Characteristic function $e^{i\mu t}(1-2i\sigma t)^{-n/2}$

r-th moment about μ

$E(X-\mu)^r$ $(2\sigma)^r \prod_{i=0}^{r-1} \frac{\Gamma(r+n/2)}{\Phi(n/2)}$

9.3. Distributional Properties

If Z is N (0,1), then Z^2 is CHI (0,1,1).

If X_i is CHI (0,1,n), i=1,2,…,k, then $\sum_{i=1}^{k} X_i$ is CHI (0,1,n), n= $\sum_{i=1}^{k} n_i$

The positive square toot of chi (0,1,n) is known as chi distribution with shape parameter n. The pdf of chi distribution is

$$\frac{x^{n-1}}{2^{(n/2)-1}\Gamma(n/2)} e^{-\frac{x^2}{2}}, x \ge 0$$

The chi distribution with n =2 is the Rayleigh distribution.

Let X_p^2 is the p-th percentile point of CHI(0,1,n) and Z_p be the p-th percentile point of N(0,1), then we have the following approximate relations

$$\chi_p^2 = \frac{1}{2}(z_p + \sqrt{3n-1})^2 \text{ for n > 100}$$

$$\chi_p^2 = n\,(1 - \frac{2}{9n} + z_p\sqrt{\frac{2}{9n}})^2 \text{ for n small n.}$$

If X and Y are independent CHI (0,1,m) and CHII (0,1,n), then

$\dfrac{X/m}{Y/n}$ is distributed as F(m,n),

Let X be a CHI (0,1,2n), then

$$P(X > x) = e^{-x/2} \sum_{i=0}^{n-1} \frac{(x/2)^i}{i!}.$$

If Y is CHI (0,1,2n+1), then

$$P(Y > y) = e^{-y/2} \sum_{i=0}^{n-2} \frac{(x/2)^{i+1}}{\Gamma(i+3/2)!}. + 2(1 - \Phi(\sqrt{x}),$$

where $\Phi(x) = \int_{-\infty}^{x} \frac{1}{\sqrt{2\pi}} e^{-\frac{u^2}{2}} du$.

9.4. Random Number Generation

Let Z'= (Z_1, Z_2, …, Z_m) and ' be a random vector whose elements are independent N (0,1) random variable and $Q = Z'AZ$ then Q is distributed as CHI (0,1,k) if A is an idempotent matrix of rank k.

Generate U_1,U_2,…, U_n random number from UN (0,1).

Set X = -2ln (U_1+U_2+…+U_n).

Then X is a CHI (0,1,2n).

Generate a random variable Z from N (0,1). Then add Z^2 to X, then it will be a CHI (0,1,2n+1).

Chapter 10

Discrete Uniform Distribution (DUN (n))

10.1. Introduction

PMF DUN(n) PX=x) $\frac{1}{n}, x = 0,1,\ldots,n-1, n \geq 1.$

Pmf of DUN (5), DUN (10) and DUN (20) are given in figures 10.1.1, 10.1.2 and 10.1.3.

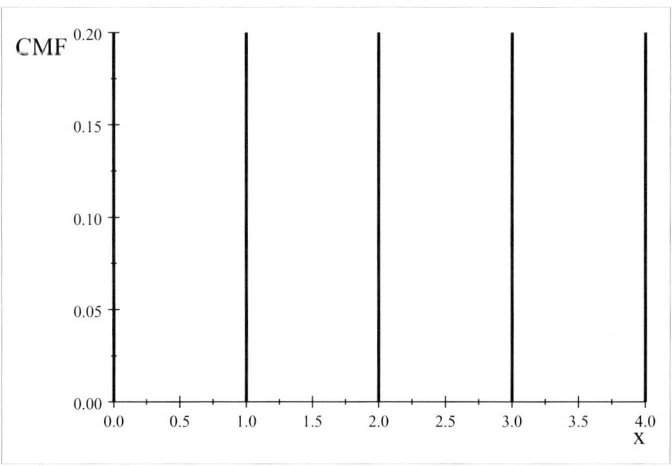

Figure 10.1.1. PMF- DUN (5).

Cdf $\frac{x+1}{n}, x = 0,1,\ldots,n-1$

CDFs of DUN (5), DUN (10) and DUN (20) are given in figures 10.1.4, 10.1.5 and 10.1.6.

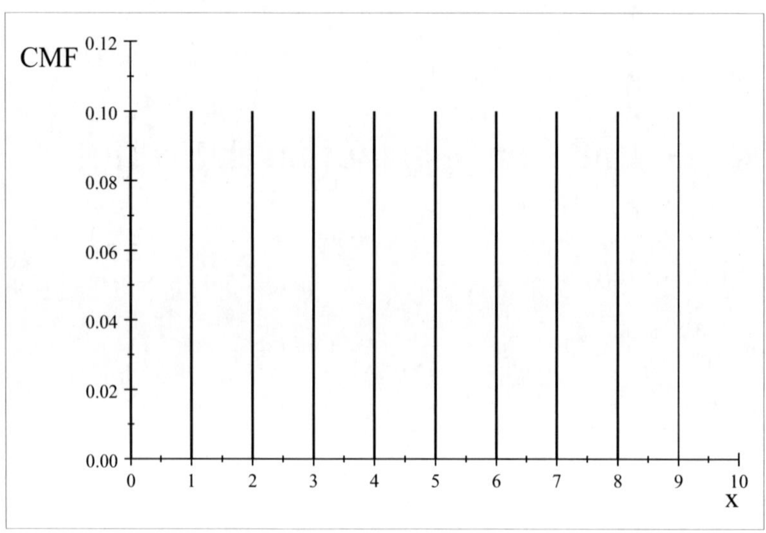

Figure 10.1.2. PDF of DUN (9).

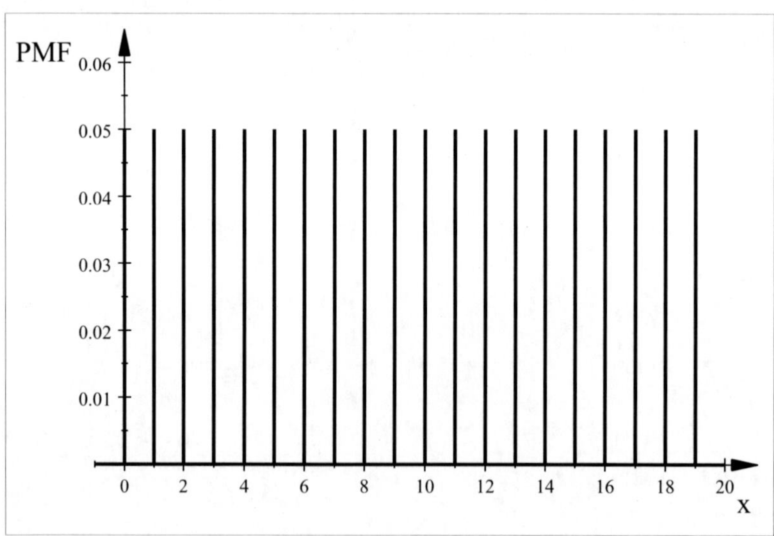

Figure 10.1.3. PDF -DUN (20).

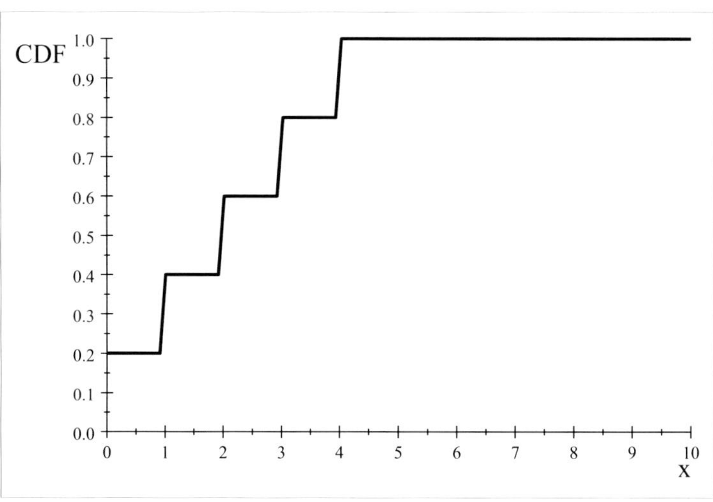

Figure 10.1.4. CDF-DUN (5).

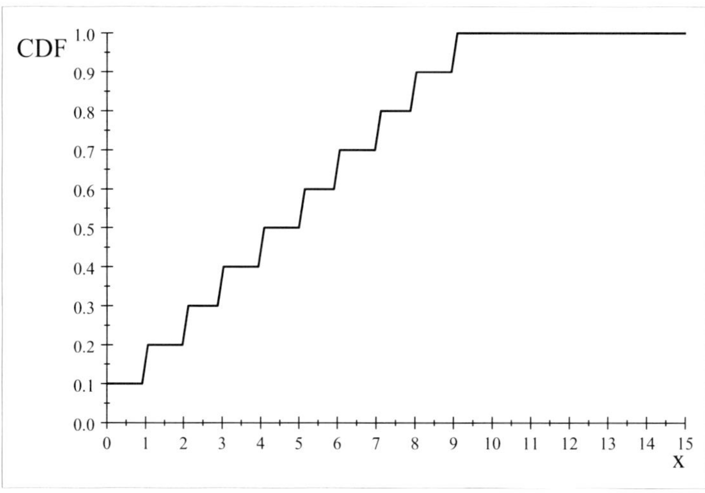

Figure 10.1.5. CDF- DUN (10).

Moments about the origin:

Mean (μ) $\frac{n-1}{2}$

$E(X^2)$		$\frac{(n-1)(2n-1)}{6}$
$E(X^3)$		$\frac{n(n-!)^2}{4}$
$E(X^4)$		$\frac{(n-1)(2n-1)(3n_2-3n-1)}{30}$
Var(X)	$E(X-\mu)^2$	$\frac{n^2-1}{12}$

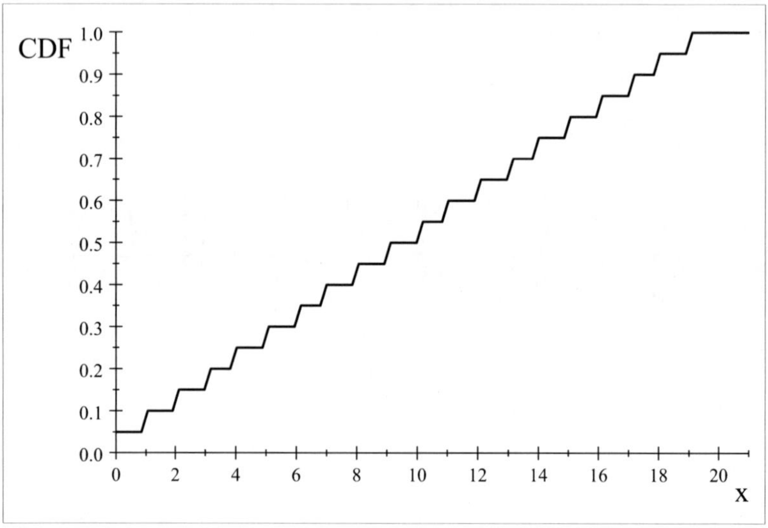

Figure 10.1.6. CDF- DUN (20).

10.2. Basic Properties

Factorial moments

$$E(X(X-1)\ldots.(X-r-1) = 0 \quad \text{if } r>n$$

$$= \frac{(n-1)\ldots(n-r)}{r+1}, \text{ if } r \leq n$$

Coefficient of variation $\sqrt{\frac{n+1}{6(n-1)}}$

Moment generating function $\dfrac{1-e^{nt}}{n(1-e^{t})}$

Characteristic function $\dfrac{1-e^{int}}{n(1-e^{it})}$

10.3. Distributional Properties

Let $\varphi_n(t)$ be the characteristic function of $Y_n = X/n$, where X is distributed as DU(n).

$$\varphi_n(t) = \frac{1}{n}(1 + e^{it/n} + e^{2it/n} + \ldots + e^{it/(n-1)/n})$$

$$= \frac{1-e^{it}}{n(1-e^{it/n})}.$$

Since $\lim_{n\to\infty} \frac{1-e^{il}}{n(1-e^{it/n})} = \frac{e^{il}-1}{it}.$

which the characteristic function of uniform, UN (0,1) distribution. Thus Y_n/ n converges in distribution to uniform distribution, UN (0,1) as n tends to ∞.

Let X_1 and X_2 be independent and are distributed as DUN(n) and DUN (m) (n>m) respectively.

Let $Y = X_1 + X_2$, then the PMF of Y is as follows

$$P(X=x) = \begin{cases} \dfrac{x+1}{mn}, 0 \le x \le m-1 \\ \dfrac{1}{n}, m \le x \le n-1 \\ \dfrac{m+n-1-x}{mn}, n \le x \le m+n-2 \end{cases}$$

If X is distributed as DUN (2), then X is distributed as Bernoulli, BER (0.5).

Chapter 11

Exponential Distribution (E(μ,σ))

11.1. Introduction

The pdf $\frac{1}{\sigma}e^{-(\frac{x-\mu}{\sigma})}, -\infty < \mu < x < \infty,\ \sigma > 0$

The exponential distribution, E (0,1) is known as standard exponential. The pdfs of E (0,1), E (0,3) and E (0,5) are given in figure 11.1.1.

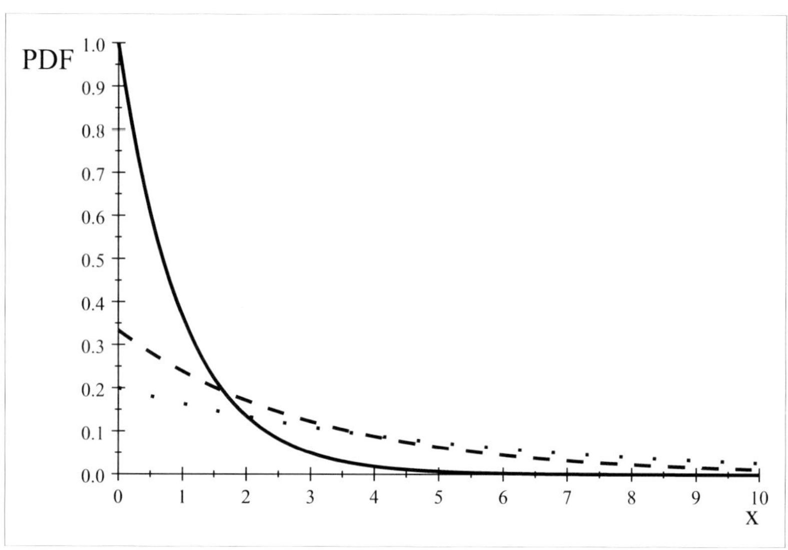

Figure 11.1.1. pdfs -E (0,1)-solid, E (0,3)-dash and E (0,5) -dots.

CDF $1-e^{-(\frac{x-\mu}{\sigma})}$.

The cdfs of E (0,1), E (0,3) and E (0,5) are given in figure 11.1.2.

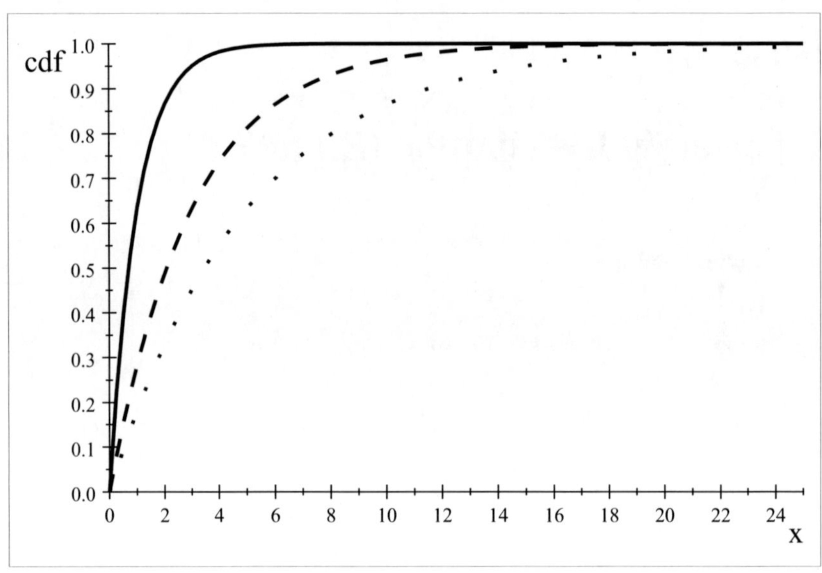

Figure 11.1.2. CDFs – E (0,1)-solid, E (0,3)-dash and E (0,5) -dots.

Mean		$\mu+\sigma$
Variance		σ^2
Coefficient of variation		$\frac{\sigma}{\mu+\sigma}$
Median		$\mu+\sigma\ln 2$
Mode		μ
Moment about zero	$E(X^k)$	$\sum_{l=0}^{k}\frac{\Gamma(k+1)}{\Gamma(j+1)}\mu^j\sigma^{k-j}$
Entropy	$1+\ln\sigma$	
Inverse distribution function	$F^{-1}(x)$	$\mu-\sigma\ln(1-x), 0<x<1$

11.2. Basic Properties

Moment generating function $\frac{e^{\mu t}}{1-\sigma t}, t < \frac{1}{\sigma}$

Characteristic function $\frac{e^{i\mu t}}{1-\sigma it}$

Hazard Rate Function $\frac{1}{\sigma}$

E(Xk) $M^{(k)}(t)|_{t=0} = \sum_{j=0}^{k} \sigma^{k-j} \frac{\Gamma(k+1)}{\Gamma(j+1)} \mu^{j}$

11.3. Distributional Properties

An important property of the exponential distribution E (0σ) is that it is memoryless. This means that if a random variable X is exponentially distributed as E (0σ), its conditional property

$$P(X > s+t \mid X > s) = P(X > t) \text{ for all s, t} \geq 0$$

Exponential distribution is the only distribution having memoryless property.

If X is distributed as E (0,ð), then

P(X>t+x) = P(X>t) P(X>x)

If X_1, X_2, …,X_n are independent and identically as E (0,ð), then $\sum_{i=1}^{n} X_i$ is distributed as Gamma, GA (n,ð).

11.4. Random Number Generation

Generate a uniform random number from a uniform U (0,1) distribution.

Set $X = \mu + \sigma \ln U$.

Then X is a random variable from E (μ, σ).

Chapter 12

F Distribution (F(m,n))

12.1. Introduction

Pdf $$\frac{\Gamma(\frac{m+n}{2})(\frac{m}{n})^{m/2}x^{(m-2)/2}}{\Gamma(\frac{m}{2})\Gamma(\frac{n}{2})(1+\frac{mx}{n})^{(m+n)/2}},$$
x≥0, m>0, n>0.

The parameters m and n are called respectively as the numerator and denominator degrees of freedoms.

The graph of pdfs F (4.4), F (12.12) and F (16.16) are given in figure 12.1.1.

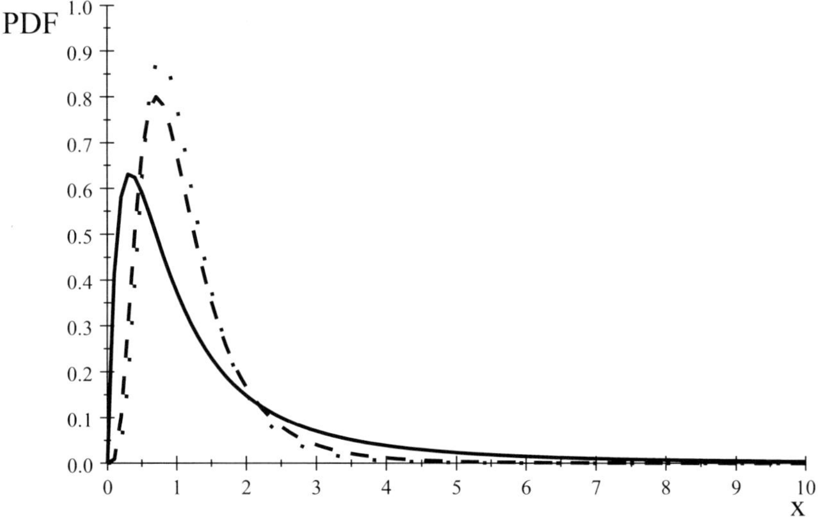

Figure 12.1.1. pdfs- F (4,4)-solid, F (12.12)- dash and F (16.16)- dots.

CDF $$I(\frac{nx}{m+n}, \frac{n}{2}, \frac{m}{2}).$$

where

$$I(x, a, b) = \frac{B(x,a,b)}{B(a,b)},$$

$$B(x,a,b) = \int_0^x u^{a-1}(1-u)^{b-1}du.$$

The cdfs of F (4,4), F (8,8) and F (16,16)-are given in figure 12.1.2.

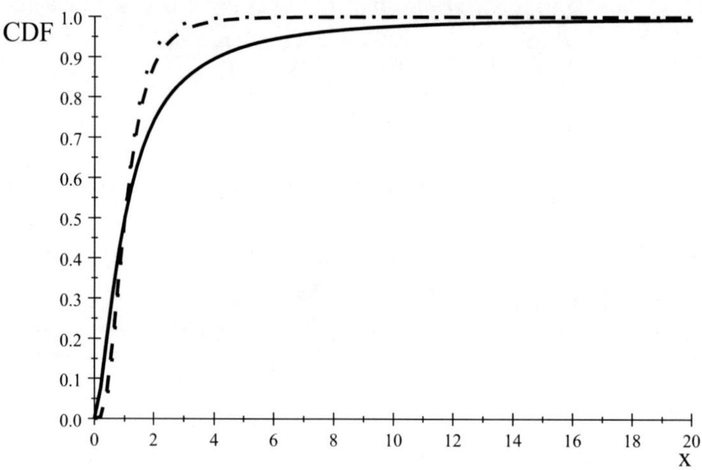

Figure 12.1.2. CDFs-F (4,4)-solid, F (8,8)-dash and F (16,16)-dots.

Mean $\frac{n}{n-2}, n > 2$

Variance $\frac{2n^2(m+n-2)}{m(n-2)^2(n-4)}, n > 4$

Coefficient of variation $\sqrt{\frac{2(m+n-2)}{m(n-4)}}, n > 4$

Moment about the origin E(X^r) $(\frac{m}{n})^r \frac{m(m+2)...(m+2r-2)}{(n-2)(n-4)....(n-2r)}$

12.2. Basic Properties

Moment generating function	Does not exist
Characteristic function	$\frac{\Gamma\left(\frac{m+n}{2}\right)}{\Gamma\left(\frac{n}{2}\right)}\Psi\left(\frac{m}{2}, 1-\frac{n}{2} \mid -\frac{n}{m}it\right)$

where $\Psi(a, b|c)$ is the confluent hypergeometric function.

12.3. Distributional Properties

If X >0 and $X_{m.n}$ is distributed as F (m,n), then $P(X_{m,n} \leq x) = P(X_{n,m} \geq 1/x)$.

If the random variable X is distributed as F (m,n), then

X^{-1} *is distributed as* F (n,m).

If X is distributed as F (m,n), then

$\frac{mX}{n+mX}$ is distributed as Beta, BE (m/2, n/2).

12.4. Random Number Generation

Take two numbers m and n.

Generate a random variable X from gamma, GA (m/2,2).
Generate a random variable Y from gamma, GA (n/2,2).
Set F = nX/mY.
Then F is a random variable from F (m,n).

Chapter 13

Gamma Distribution (GA(*a*,b))

13.1. Introduction

Pdf $\frac{1}{\Gamma(a)b^{a}}x^{a-1}e^{-x/b}, x \geq 0, a > 0, b > 0$’

The pdfs of GA (2,1), GA (6.1) and GA (10,1) are given in figure 13.1.1.

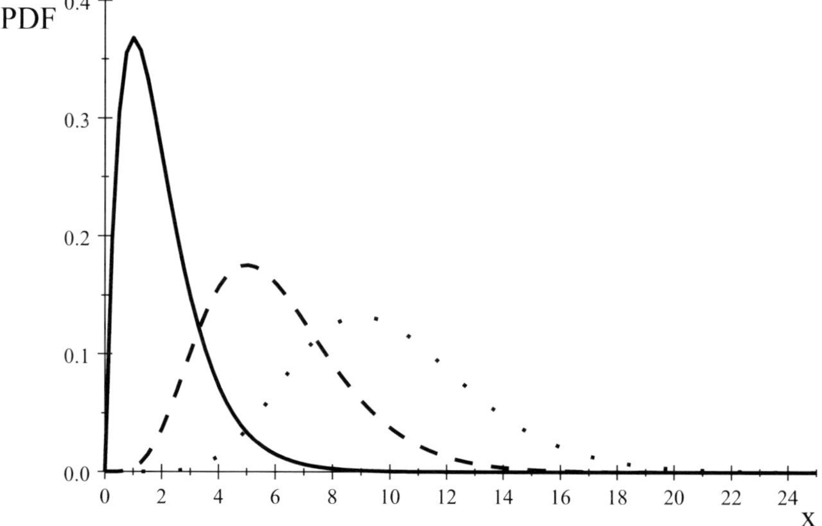

Figure 13.1.1. PDFs- GA (2,1)-solid, GA (6,1)-dash and GA (10,1)-dots.

Cdf $\frac{1}{\Gamma(a)}\gamma(\frac{x}{b}, a)$

Figure 13.1.2. gives the cdfs of GA (2,1), GA (6.1) and GA (10,1).

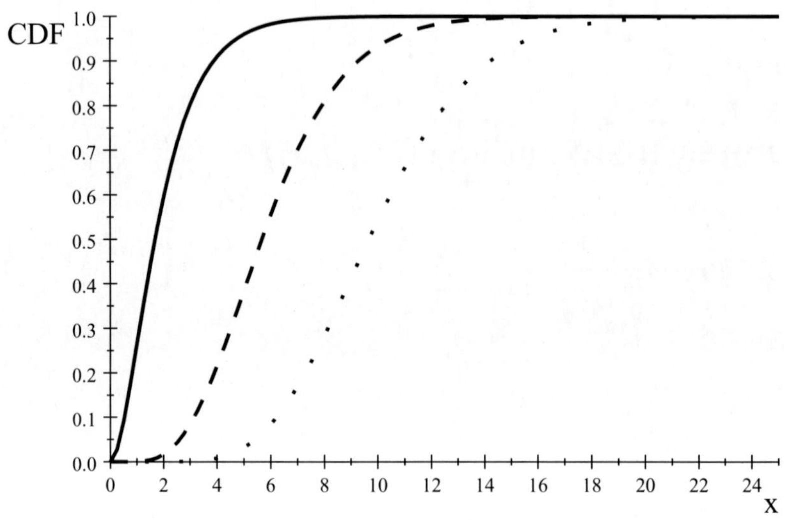

Figure 13.1.2. Cdfs -GA (2,1),-solid. GA (6.1)-dash and GA (10,1)- dots.

Mean	ab
Variance	ab^2
Coefficient of variation	$\frac{1}{\sqrt{a}}$

13.2. Basic Properties

Moment generating function	$(1-bt)^{-a}, t < \frac{1}{b}$
Characteristic function	$(1-ibt)^{-a}$
Moment about the origin E(X^r)	$\frac{\Gamma(a+r)}{\Gamma(a)} b^k$

The hazard rate function of GA (1,1) is equal to 1 and the hazard rate function of of GA (a,b) tends to 1/b as x tends to infinity.

13.3. Distributional Properties

If $X_1, X_2 \ldots, X_n$ are n independent exponential random variable from ,E (0,1), then

$\sum_{i=1}^{n} X_i$ is distributed as GA (n,1).

If $X_1, X_2 \ldots, X_n$ are n independently distributed as GA (a_i,b), then

$\sum_{i=1}^{n} X_i$ is distributed as GA $(\sum_{i=1}^{n} a_i b)$

If a=n/2 and b=2, then GA(n/2,2) is distributed as CHI (n).
If X_1 and X_2 are independently distributed as GA (n,1), then

$\sqrt{\frac{n}{2}}(\frac{X_1 - X_2}{\sqrt{X_1 X_2}})$ is distributed as ST (2n).

13.4. Random Number Generation

Generate n independent uniform, UN (),1) random variables, $U_1, U_2, \ldots, U_n$.

Set $X = -\sum_{j=1}^{n} In\ U_j$
Then X is GA (n,1) random variable.

Chapter 14

Geometric Distribution (GE(p))

14.1. Introduction

PMF $p(1\text{-p})^x, x = 0,1,\ldots,0 < p < 1$

The pmfs of GE(0.5),GE(.5) and GE(.8) are given in figures 14.1.1 and 14.1.2.

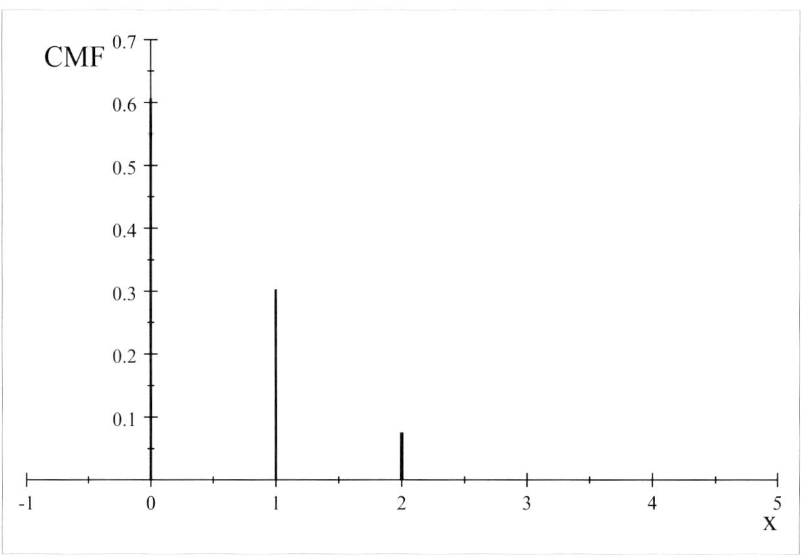

Figure 14.1.1. PMF- GE (0.5).

CDF 1-$(1-p)^{x+1}$, x=0.1.2.

CDFs of GE (0.5) and GE (0.8) are given in figures 14.1.3 and 14.1.4.

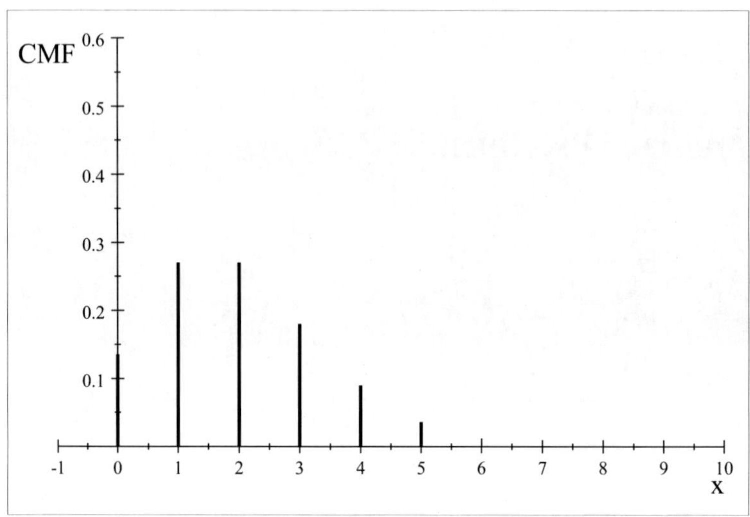

Figure 14.1.2. PMF-GE (0.8).

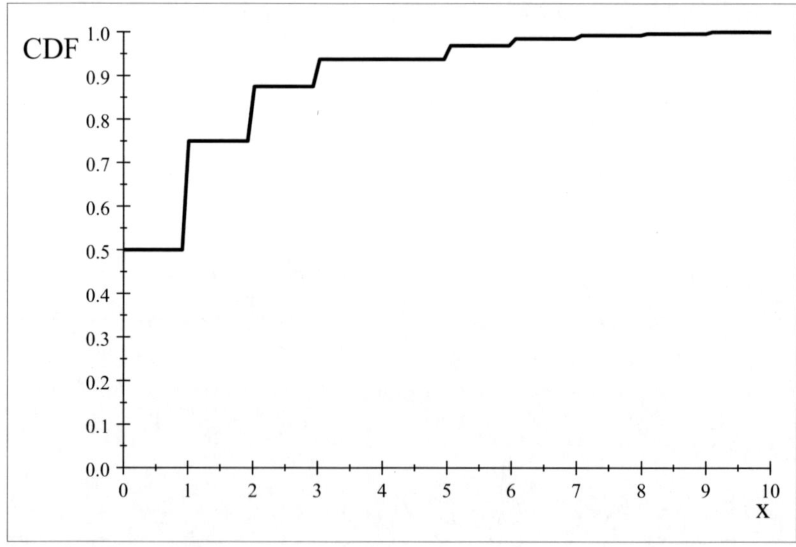

Figure 14.1.3. PMF-GE (0.6).

Mean $(1-p)/p$

Variance $(1\text{-}p)/p^2$

Moment about the mean $\mu_r = E(X - \frac{q}{p})^r, q = 1 - p$

$$\mu_{r+1} = (1 - p)\frac{d\mu_r}{dq} + \frac{r}{p^2}\mu_{r,\mu_0=1,\mu_1=0},$$

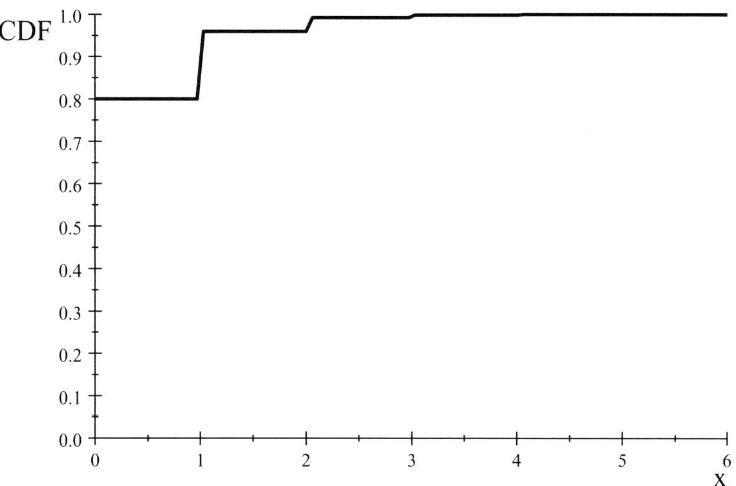

Figure 14.1.4. CDF -GE (0.8).

In particular

$$\mu_3 = q(1+q)/p^3$$
$$\mu_4 = 9q^2/p^4 + q/p^2$$

14.2. Basic Properties

Moment generating function $p\,(1\text{-}qe^t)^{-1}$

Characteristic function $p\,(1\text{-}qe^{it})^{-1}$

Probability generating function $p(1\text{-}qt)^{-1}$

Let y be the expected number of failures, then

$$E(Y) = p\sum_{k=0}^{\infty} k(1-p)^k$$
$$= p(1-p)\sum_{k=0}^{\infty} k(1-p)^{k-1}$$
$$= p(1-p)(-\sum_{k=0}^{\infty} \frac{d}{dp}(1-p)^{k)}$$
$$= p(1-p)[\frac{d}{dp}(-\sum_{k=0}^{\infty}(1-p)^{k)}$$
$$= p(1-p))\frac{d}{dp}(--\frac{1}{p})$$
$$= \frac{1-p}{p}$$

14.3. Distributional Properties

If X_i, i=1,2,…, m are independently distributed GE (p), them

$\sum_{i=1}^{m} X_i$ is distributed as negative binomial, NB (m,p).

If X is distributed as G(p), then $P(X>n+m|X>m) = P(X>m)$.
Let X_1 and X_2 be independent and identically distributed as GE (p), then

$$P(X_1=x|\ X_1+X_2=y) = \frac{1}{1+y},\ y = 0,1,\ldots$$

$$P(X \geq k+1) = q^{k+1}$$

14.4. Random Number Generation

Generate a random number U from Uniform distribution. Select a p. $0<p<1$.

Set $X = \left|\frac{lnU}{\ln(1-p)}\right|$

X is a random variable from GE (p).

Chapter 15

Gumbel Distribution (GU (μ,σ))

15.1. Introduction

Pdf $\frac{1}{\sigma}\exp\left(-\frac{x-\mu}{\sigma}\right)\exp\left(-\exp\left(-\frac{x-\mu}{\sigma}\right)\right)$.$-\infty < \mu < x < \infty, \sigma > 0$

The graph of GU (0,1/4), GU (0,1) and GU (0,3) are given in figure 15.1.1.

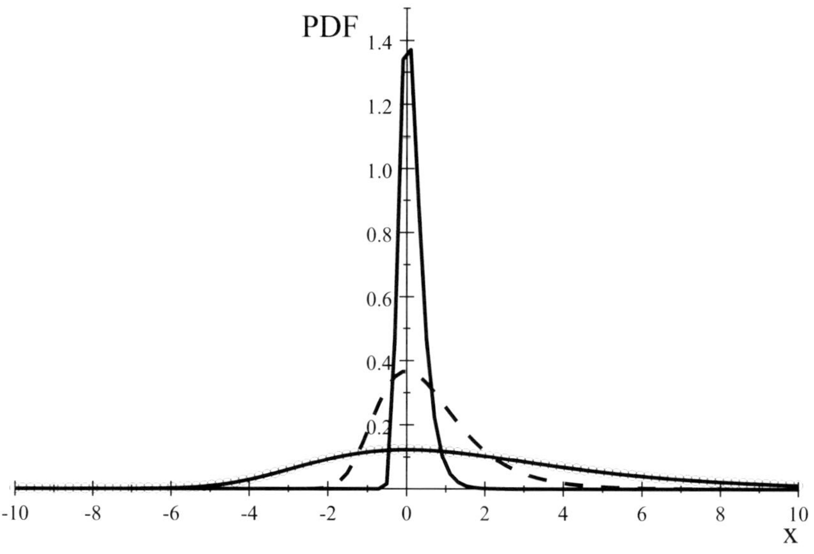

Figure 15.1.1. CDFs- GU (0.1/4\)-sold, GU (0,1)-dash and GU (0,3)-dotsdash.

Basic Properties

Cdf $e^{-e^{-(\frac{x-\mu}{\sigma})}}$

The cdfs of GU (0,0,25), GU (0,1) and GU (0,3) are given in figure 15,1.2.

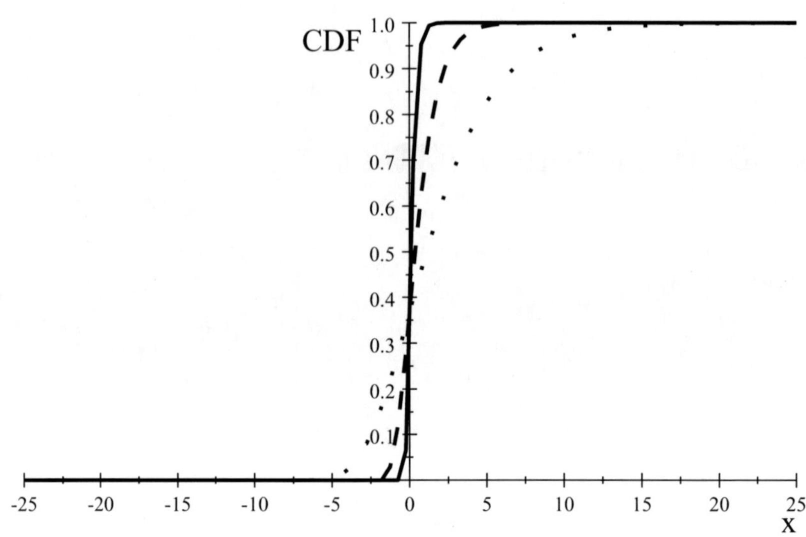

Figure 15.1.2. CDFs- GU (0,0.25)-solid, GU (0,1)-dash and GU (0,3)-dots.

Table 15.1. Percentile points

P	GU(0,1/2)	GU (0,1)	GU (0,2)
0.05	-0.5486	-1.0972	-2.1944
0.1	-0.4170	-0.8340	-1.6681
0.2	-0.2379	-0.4759	-0.9518
0.3	-0.0928	-0.1856	-0.3713
0.4	-0.0437	=0.0874	0.1748
0.5	0.18833	0.3665	0.7130
0.6	0.3359	0.8717	1.3435
0.7	0.5155	0.0309	2.0619
0.8	0.7500	1.4999	2.0000
0.9	1,1252	2.2504	4.5007
0.95	1.4851	2.9702	5.9404

Mean

$\mu + \gamma\sigma,\ \gamma$ is Euler constant

Median $\mu - \sigma \ln(n2)$

Mode μ

Variance	$\frac{\pi^2\sigma^2}{6}$
The points of inflection	$\mu \pm \sigma\, ln\ (\frac{1+\sqrt{5}}{2})$
Coefficient of variation	$\frac{\pi\sigma}{\sqrt{6}(\mu+\gamma\sigma)}$
Entropy	$ln\,\sigma + \gamma + 1$
First quartile	$\mu - 0.3266\sigma$
Third quartile	$\mu + 1.2459\sigma$

15.2. Basic Properties

Moment generating function	$e^{\mu t}\Gamma(1-\sigma t)$
Characteristic function	$e^{\mu it}\Gamma(1-\sigma it)$
Hazard rate	$\frac{\exp\left(-\frac{x-\mu}{\sigma}\right)\exp\left(-\exp\left(-\frac{x-\mu}{\sigma}\right)\right)}{\sigma(1-\exp\left(-\exp\left(-\frac{x-\mu}{\sigma}\right)\right))}$

15.3. Distributional Properties

The tail of the pdf of Gumbel distribution GU (0, σ) as X tends to infinity coincides with the tail of the pdf of exponential distribution., E (0, σ).

If X is exponentially distributed as E (0,1}, then $\mu - \sigma \ln X$ is distributed as GU(μ,σ).

If X and Y are independently distributed as GU (μ,σ), then X-Y is distributed as LO ($0,\sigma$).

If X has GU (0,1), the e^X has a Weibull distribution, WE (0,1),

The largest order statistics from many continuous distributions with suitable normalized constants converse in distribution to GU (0,1), i.e., for many distributions

$$lim_{n\to\infty} P\left(X_{n,n} \leq \alpha_n + \beta_n x\right) = e^{-e^{-x}}$$

where $X_{n,n} = \max(X_1, X_2, \ldots, X_n)$ of independent and identically distributes random variables

The following table gives α_n and β_n for some selected distribution.

Table 15.2. Normalizing constants α_n *and* β_n

Distribution	pdf	α_n	β_n
Beta	$cx^{\alpha-1}(1-x)^{\beta-1}$ $c = \frac{\Gamma(\alpha+\beta)}{\Gamma(\alpha)\Gamma(\beta)}$	1	$(\frac{\beta}{nc})^{1.\phi}$
Cauchy	$\frac{1}{\pi(1+x)^2}$	0	$\frac{n}{\pi}$
Exponential	$\sigma e^{-\sigma x}$	$\frac{1}{\sigma}\ln n$	$\frac{1}{\sigma}$
Gamma	$\frac{x^{\phi-1}e^{-x}}{\Gamma(\alpha)}$	$\ln n - \ln\Gamma(\alpha)$ $+(\alpha-1)\ln\ln n$	1
Laplace	$\frac{1}{2}e^{-\lvert x\rvert}$	ln(n/2)	1
Logistic	$\frac{e^x}{(1+e^x)^2}$	lnn	1
normal	$\frac{1}{\sqrt{2\pi}}e^{-\frac{x^2}{2}}$	$\frac{1}{\beta_n} - \frac{\beta_n D_n}{2}$ $D_n = \ln\ln n + in4\pi$ $\beta_n = (2\ln n)^{-1/2}$	$(2\ln n)^{-1/2}$
Raleigh	$2xe^{-x^2}$	$(\ln 2)^{1/2}$	$\frac{1}{2}(\ln 2)^{-1/2}$

15.4. Random Number Generation

Generate a random variable U from a uniform distribution, UN (0,1).

Set X =μ-σ ln (-ln U). Then X is GU (μ,σ).

Chapter 16

Hypergeometric Distribution (HG(*a*,b,n))

16.1. Introduction

PMF P(X=m) $\frac{\binom{a}{m}\binom{b}{n-m}}{\binom{a+b}{n}}$, n = 1,2,…,*a*+b

m={max(0,n-b),…,min(n,*a*)}

The PMFs of HG (6,6,4). HG (8,8,5) are given in figures 16.1.1 and, 16.1.2.

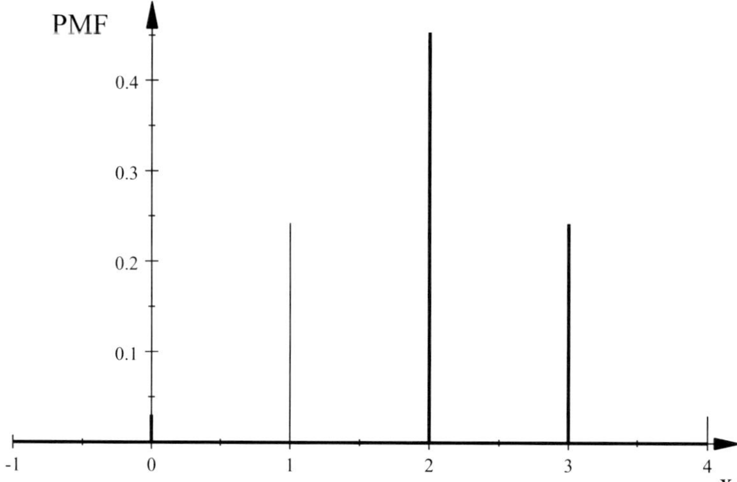

Figure 16.1.1. PMF- HG (6,6,4).

The cdfs of HG (6,6,4) and HG (8,8,5) are given in figures 16.1.3 and 16.1.4.

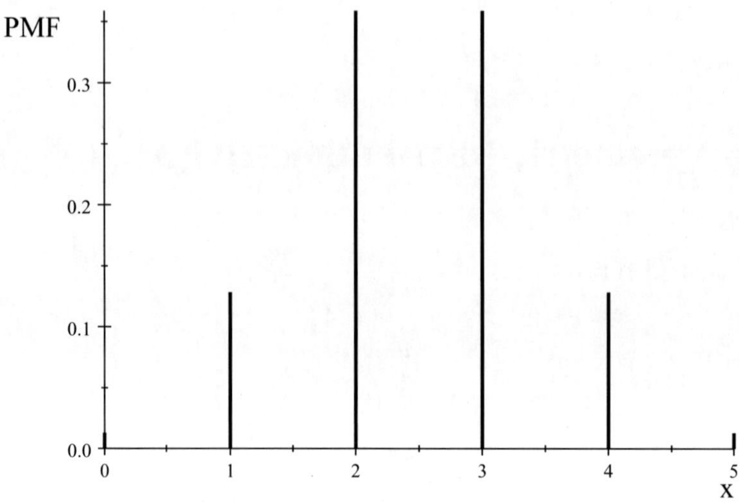

Figure 16.1.2. PMF-HG (8,8,5).

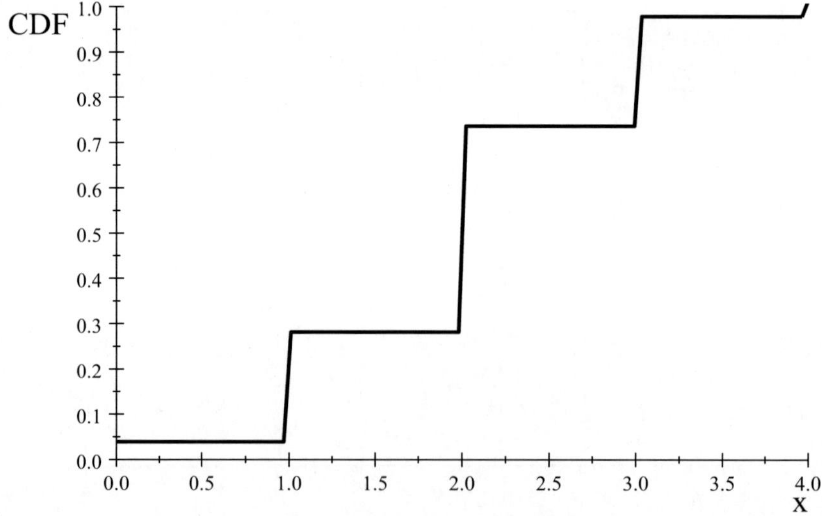

Figure 16.1.3. CDF- HG (6,6,4).

Mean $\qquad \dfrac{na}{a+b}$

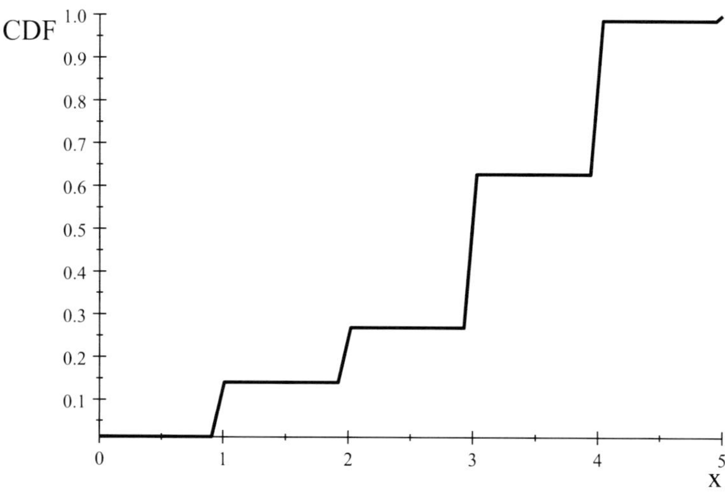

Figure 16.1.4. CDF -HG (8.8.5\).

Variance $\frac{nab(a+b-n)}{(a+b)^2(a+b-1)}$

Coefficient of variation $\sqrt{\frac{b(a+b-n)}{na(a+b-1)}}$

Factorial moments μ_f^k

$$\mu_f^k \quad = EX(X-1)(x_2 \quad ...(X-k+1)$$

$$\mu_f^k \quad = \sum_{m=k}^{n} m(m-1)(m-2)(m-k+1)\ \frac{\binom{a}{m}\binom{b}{n-m}}{\binom{a+b}{n}}$$

$$= \frac{a!b!n!(a+b-n)!}{(a+b)!(n-k)!} \sum_{m=0}^{n-k} \frac{1}{m!(n-k-m!(n-k-m)!(b-n+k+m)!}$$

$$= \frac{a!n!(a+b-k)!}{(a+b)!(a-k)!(n-k)!},\ \text{k}\leq \min\,(a,n)$$

$$= 0,\ \text{k}\geq \min\,(a,n)$$

$$\mu_f^1 = \frac{na}{b}$$

$$\mu_f^2 = \frac{n(n-1)a(a-1)}{(a=b)(a+b-1)}$$

16.2. Basic Properties

Moment generating function

$$\frac{{}_2F_1(-n,-a;b-a+1:e^t)}{{}_2F_1(-n,-a;b-a+1:1)}$$

where ${}_2F_1(\alpha,\beta;\gamma:x)$ is the Gaussian Hypergeometric function given by

$${}_2F_1(\alpha,\beta;\gamma:x)=1+\frac{\alpha\beta}{\gamma}\frac{x}{1!}+\frac{\alpha(\alpha+1)\beta(\beta+1)}{\gamma(\gamma+1)}\frac{x^2}{2!}+.....$$

Characteristic function $\frac{{}_2F_1(-n,-a;b-a+1:e^{it})}{{}_2F_1(-n,-a;b-a+1:1)}$

Probability generating function $\frac{{}_2F_1(-n,-a;b-n+1;t)}{{}_2F_1(-n,-a;b-n+1;1)}$

HG(a,b,1) is Bernoulli, BE($a,(a+b)$).

For HG (a,b,n), P(X=x+1) = $\frac{(n-x)(a-x)}{(x+1)(b-n+x+1)}p(X=x)$.

16.3. Distributional Properties

For HG (m,m,r),

$$P(X=x)=\frac{\binom{m}{x}\binom{m}{r-x}}{\binom{2m}{r}}=P(X=r-x)$$

If m $\to\infty$, then P(X=0) = $\frac{1}{2^r}$.

As m $\to\infty, H(m,m,n)\to BN(n,\frac{1}{2})$

As m $\to\infty, H(m,km,n)\to BN(n,1/(k+1))$

If X is distributed as HG (1,a,b) , then X is distributed as BE ($\frac{a}{a+b}$)

If X is distributed as HG (m,N,n) such that n and m tend to infinity and m/N =p and for p not close to 0 or 1, then

$$P(X\leq k) =\cong \Phi(\frac{k-np}{\sqrt{np(1-p)}})$$

where Φ (.) is the cdf of standard normal (0,1).

Chapter 17

Inverse Gaussian Distribution ($\mathrm{IG}(\mu,\lambda)$)

17.1. Introduction

Pdf $$\left(\frac{\lambda}{2\pi x^3}\right)^{1/2} e^{-\frac{\lambda}{2x}\left(\frac{x-\mu}{\mu}\right)^2}, x>0, \lambda, \mu > 0$$

The pdfs of IG (1,1), IG (2,2) and IG (5,5) are given in figure 17.1.1.

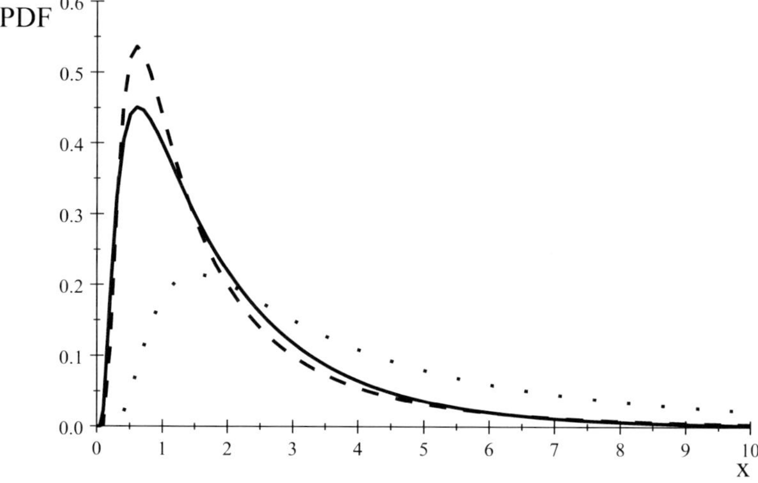

Figure 17.1.1. Pdfs – IG (1,1)-solid, IG (2,2)-dash and IG (5,5) -dots.

CDF $$\Phi\left(\sqrt{\tfrac{\lambda}{x}}\left(\tfrac{x}{\mu}-1\right)\right)+e^{2\lambda/\mu}\,\Phi\left(\sqrt{\tfrac{\lambda}{x}}\left(1+\tfrac{x}{\mu}\right)\right)$$

The cdfs of IG (1,1), IG (2,2) and IG (3,3) are given in figure 17.1.2.

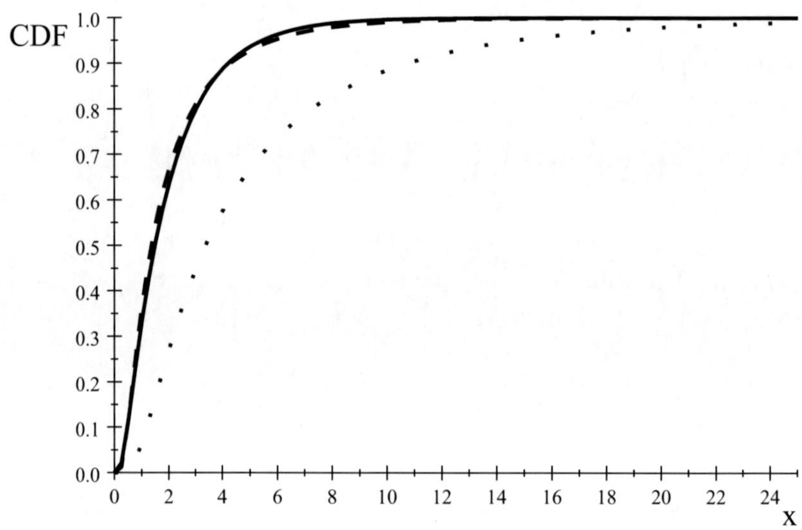

Figure 17.1.2. Cdfs -I G (1,1)-solid., IG (2,2) -dash and IG (3,3)-dots.

Mean	μ
Variance	$\frac{\mu^3}{\lambda}$
Coefficient of variation	$\sqrt{\frac{\mu}{\lambda}}$
Moment about the origin $E(X^{k})$	$\mu^k \sum_{j=0}^{k-1} \frac{(k-1+j)!}{(k-1-j)!}(\frac{\mu}{2\lambda})^j, k \geq 2$
$E(X^2)$	$\frac{\mu^2(\mu+\lambda)}{\lambda}$
$E(X^3)$	$\frac{\mu^3(\mu^3+2\mu\lambda+\lambda^2)}{\lambda^2}$
$E(X^{m+1})$	$\frac{(2m-1)\mu^2}{\lambda} E(X^m) + \mu^2 E(X^{m-1})$

17.2. Basic Properties

Moment generating function

$$e^{[\frac{\lambda}{\mu}(1-(1-\frac{2\mu^2 t}{\lambda})^{\frac{1}{2}})]}$$

Characteristic function

$$e^{[\frac{\lambda}{\mu}(1-(1-\frac{2i\mu^2 t}{\lambda})^{\frac{1}{2}})]}$$

Cdf $$\Phi(\sqrt{\frac{\lambda}{x}}(\frac{x}{\mu}-1)) + e^{2\lambda/\mu}\Phi(-\sqrt{\frac{\lambda}{x}}(\frac{x}{\mu}+1))$$

where $\Phi(x) = \int_{-\infty}^{x} \frac{1}{\sqrt{2\pi}} e^{-\frac{u^2}{2}} du$

Hazard rate function(HRT) $$\frac{(\frac{\lambda}{2\pi x^3})^{1/2} e^{-\frac{\lambda}{2x}(\frac{x-\mu}{\mu})^2}}{1-(\Phi(\sqrt{\frac{\lambda}{x}}(\frac{x}{\mu}-1)) + e^{2\lambda/\mu}\Phi(-\sqrt{\frac{\lambda}{x}}(\frac{x}{\mu}+1)))}$$

for $x > 0, \lambda, \mu > 0$

17.3. Distributional Properties

If X is distributed as IG (λ, μ),then $\alpha X$is distributed as IG $(\alpha\lambda, \alpha\mu)$.

If X is distributed as IG $(\lambda,1)$,then the distribution of X is known as Wald distribution.

If $X_1, X_{2,\dots} X_n$ are independent and identically distributed as IG(μ, λ), $then\ \sum_{i=1}^{n} X_i\ is\ ditributed\ as$ IG (n$\mu.\, n\lambda$)

If X is distributed as IG (μ, λ), $then\ \frac{\lambda(x-\kappa)^2}{\mu^2 X}\ is\ distributed\ as\ chi-sqrare\ with\ 1\ degree\ of\ freedom.$

17.4. Random Number Generation

For given μ, λ, generate U from uniform distribution, UN (0,1) and Z from normal distribution, N (0,1)

Set

$V=Z^2$,
$D = \lambda/\mu$.
$Y = 1 - 0.5(\sqrt{U^2 + 4DV} - V)/D$
$X = Y\mu$

If $(1+Y)U > 1$, Set $X = 1/Y\mu$.
$\mu IG\ (\lambda$

Then X is a random number from IG (μ, σ)

Chapter 18

Laplace Distribution (LA (μ, σ))

18.1. Introduction

Pdf $\frac{1}{2\sigma} e^{-|\frac{x-\mu}{\sigma}|}. -\infty < x, \mu < \infty. \sigma > 0.$ -

LA (0,1) is known as standard Laplace distribution.
The pdfs of LA (0,1), LA (0,21) and LA (0,3) are given in figures 18.1.1.

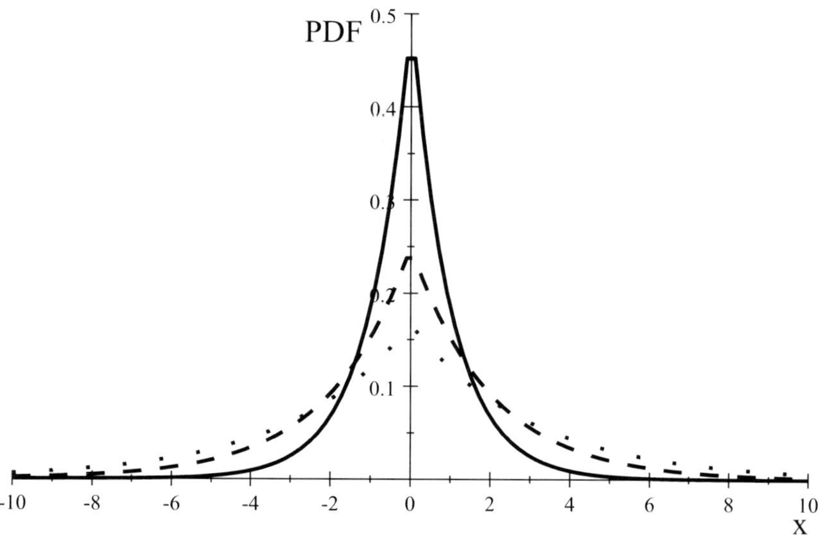

Figure 18.1.1. pdfs- LA (0,1)=solid, LA (0,2)-dash and LA (0,3) -dots.

Cdf $\frac{1}{2} e^{\frac{x-\mu}{\sigma}}$ if x<μ

$1 \text{--} \frac{1}{2} e^{-\frac{x-\mu}{\sigma}}$

The cdfs of LA (0,1), LA (0,2) and LA (0,5) are given in figure 18.1.2.

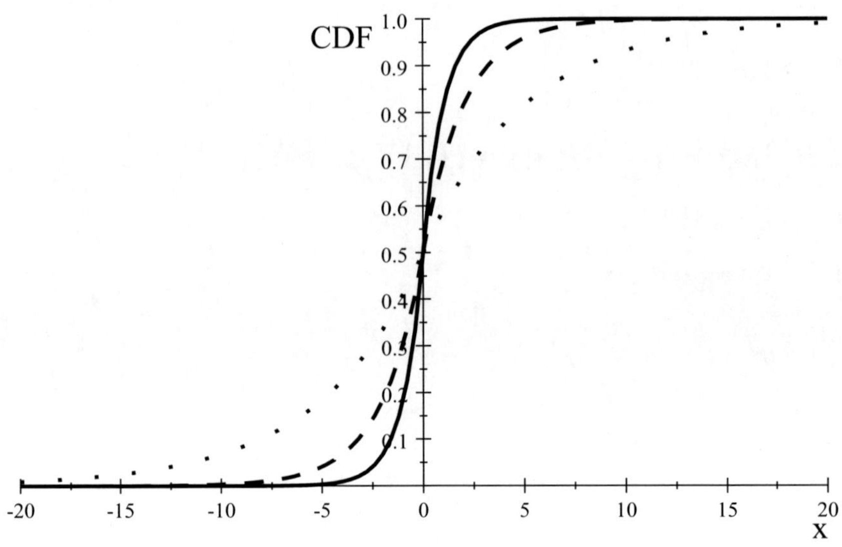

Figure 18.1.2. Cdfs - LA (0,1)-solid, LA (0,2)-dash and LA (0,5)-dots.

Mean	μ
Median	μ
Mode	μ
Variance	$2\sigma^2$
Coefficient of variation	$\frac{\sigma\sqrt{2}}{\mu}$
Entropy	$1+\ln(2\sigma)$

18.2. Basic Properties

Moment generating function	$\frac{e^{\mu t}}{1-\sigma^2 t^2}$
Characteristic function	$\frac{e^{i\mu t}}{1+\sigma^2 t^2}$

The r-th moment about μ,

E(X-μ) $= 0$ if r is odd,

$= r!\sigma^r$ if r is even.

First quartile $\mu - \frac{\ln 2}{\sqrt{2}}\sigma$

Third quartile $\mu + \frac{\ln 2}{\sqrt{2}}\sigma$

Hazard rate function $\frac{e^x}{2-e^x}, x<0$

$1, \quad x \geq o$

The hazard rate increases for x<0 and is constant for $x \geq 0$.

18.3. Distributional Properties

If Y has a standard exponential distribution, E (0,1) and Z has a standard normal distribution, N (0,2), then $\sqrt{Y}Z$ is distributed as LA (0,1).

If X is LA (0,1), then

$$X \underline{\underline{d}}\, E_1 - E_2$$

where E_1 and E_2 are independently distributed as standard exponential, E(0,1).

Since exponential distribution is infinitely divisible, therefore the Laplace random variable ` X is infinitely divisible.

If $X_i, i = 1,2, are$ distributed as $LA\ (0, \sigma_i)$. I=1.2 and $X_1$and X_2 are independent, then $X_1 + X_2$ and $X_1 - X_2$ are identically distributed. If $\sigma_1 = \sigma_2 = \sigma$, then their common pdf, f(x) is

$$F(x) = \frac{1}{4\sigma}(1+\frac{|x|}{\sigma})\exp(-\frac{|x|}{\sigma}).$$

For LA (0,1), then 2|X| is distributed as chi-square with 2 degrees of freedom

If $X_1, X_2, \ldots, X_n$ are independently distributes as LA (0,1), then

$2\sum_{i=1}^{n} |X_i|$ is distributed as Chi-square with 2n degrees of freedom

$\frac{|X_1|}{|X_2|}$ is distributed as $F_{2.2}$

If U_1 and U_2 are independently distributed as uniform UN (0,1) random variables, then

$|\ln(\frac{U_1}{U_2})|$ is distributed as LA (0,2)

18.4. Random Number Generation

Generate a uniform random number U from a uniform, UN (0,1) distribution.

If U>1/2, return $X = \mu - \sigma \ln(2(1-U))$
If U <1/2, return $X = \mu + \sigma \ln(2U)$

X is a random number from LA $(\mu . \sigma)$

Chapter 19

Logarithmic Series Distribution (LS (λ))

19.1. Introduction

PMF $\theta \frac{\lambda^x}{x}, x = 1,2,\ldots,0 < \lambda < 1, \theta = -\frac{1}{ln(1-\lambda)}$

The PMF of LS (0.5) and LS (0.7) are given in figures 19.1.1 and 19.1.2 and 15.2.

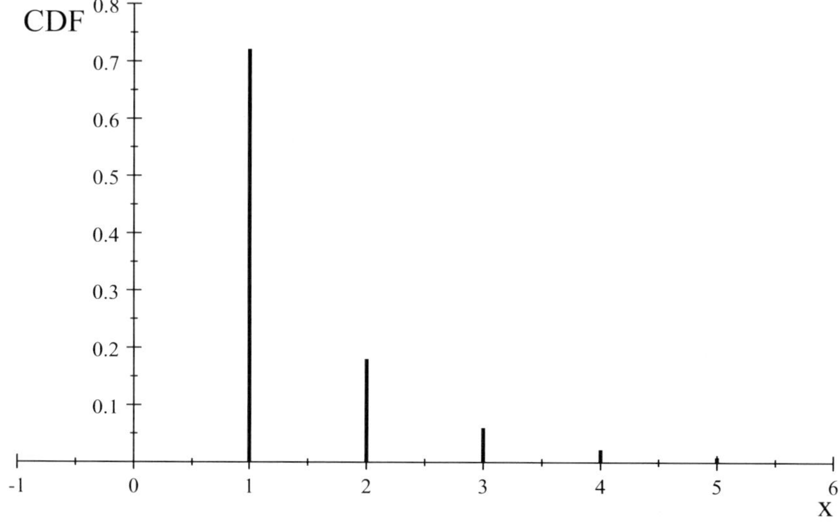

Figure 19.1.1. PMF-LS (0.5).

CDF $1+\frac{B(\lambda,x+1,0)}{\ln(1-p)}$

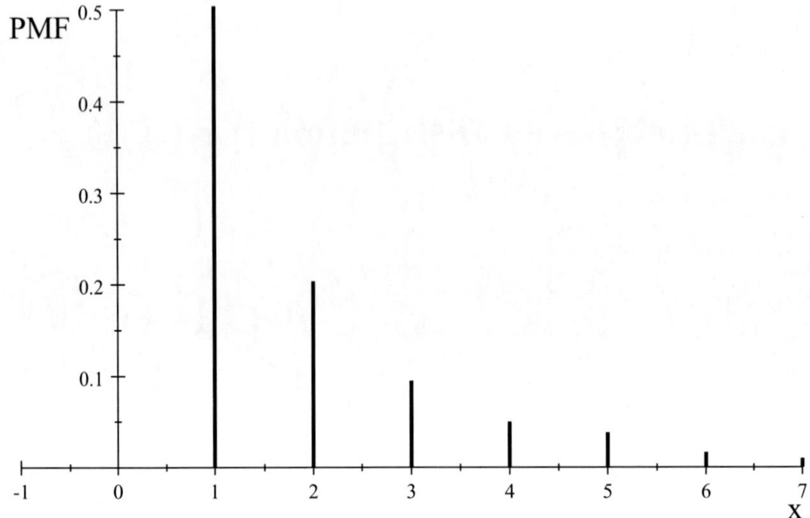

Figure 19.1.2. PMF- LS (0.7).

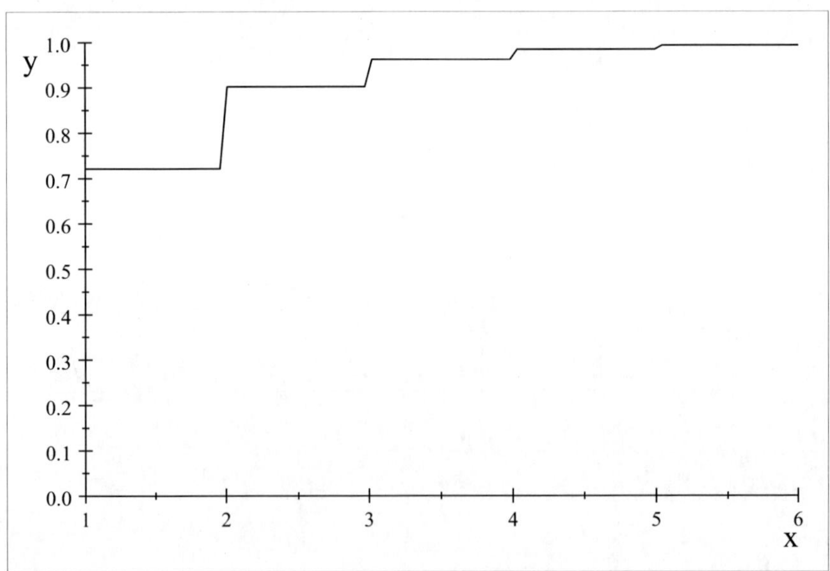

Figure 19.1.3. CDF- LS (0.5).

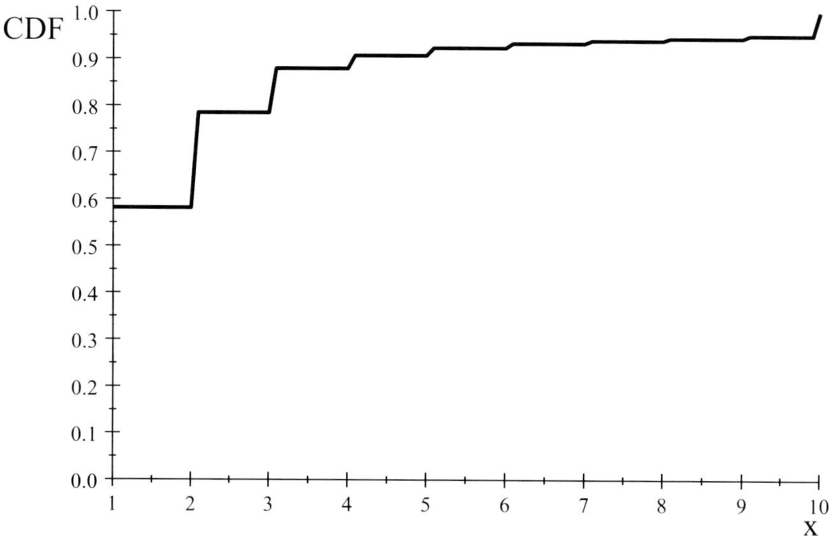

Figure 19.1.4. CDF- LS (0.7).

Mean	$\frac{\theta\lambda}{1-\lambda}, \quad \theta = \frac{1}{-\ln(1-\lambda)}$
Second moment	$\frac{\theta\lambda}{(1-\lambda)^2}$
Third moment	$\frac{\theta\lambda(1+\lambda)}{(1-\lambda)^3}$
Fourth moment	$\frac{\theta\lambda(1+4\lambda+\lambda^2)}{1-\lambda)^4}$
Variance	$\frac{\lambda\theta(1-\theta\lambda)}{(1-\lambda)^2}$
Coefficient of variation	$\sqrt{\frac{1-\theta\lambda}{\theta\lambda}}$

19.2. Basic Properties

Moment generating function $-\theta\, ln(1-\lambda e^{t}), t < -ln\,\lambda$

Characteristic function $-\theta\, ln(1-\lambda e^{it})$

19.3. Distributional Properties

If X_1 ,X_2, …, X_5 are identically distributed as LS (λ), then the PMF of $Y = \sum_{k=1}^{n} X_k$ is

$$P(Y = m) = \frac{n!|\, S_m^{(n)} \,|\, \lambda^k \theta^n}{m!}, m = n, n+1, ...$$

where is the $S_m^{(n)}$ Stirling number of the first kind.

$$\frac{P(X = x+1)}{P(X = x)} = \frac{\lambda x}{x+1}, x = 1,2,....$$

Chapter 20

Logistics Distribution (LO (μ, σ))

20.1. Introduction

Pdf $$\frac{1}{\sigma}\frac{e^{-\frac{x-\mu}{\sigma}}}{(1+e^{-\frac{x-\mu}{\sigma}})^2}, -\infty < \mu < x < \infty, \sigma > 0.$$

Figure 20.1.1 shows the PDFs for LO (0,0.5), LO (0,1), and LO (0,2).

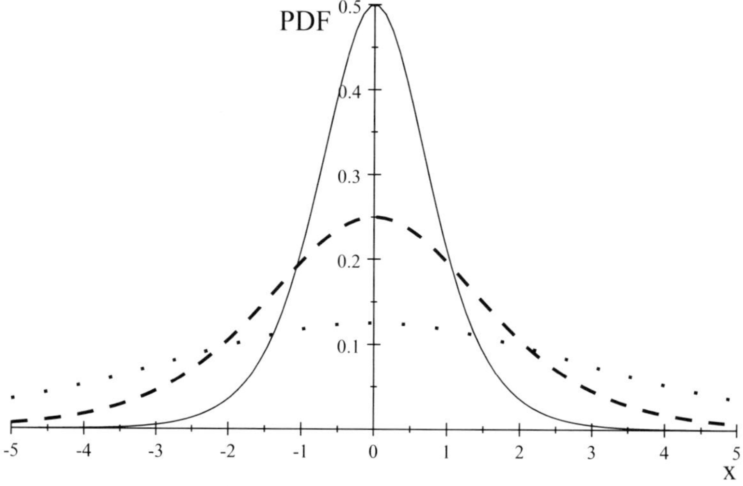

Figure 20.1.1. Pdfs- LO (0,0.5),-solid, LO (0,1)-dash and LO (0,2) -dots.

CDF $$(1+\exp(\frac{x-\mu}{\sigma}))^{-1} = \text{-}\frac{1}{2}+\frac{1}{2}\tanh(\frac{x-\mu}{2\sigma})$$

The cdfs of LO (0,0.5), LO (0,1) and LO (0,2) are given in figure 20.1.2.

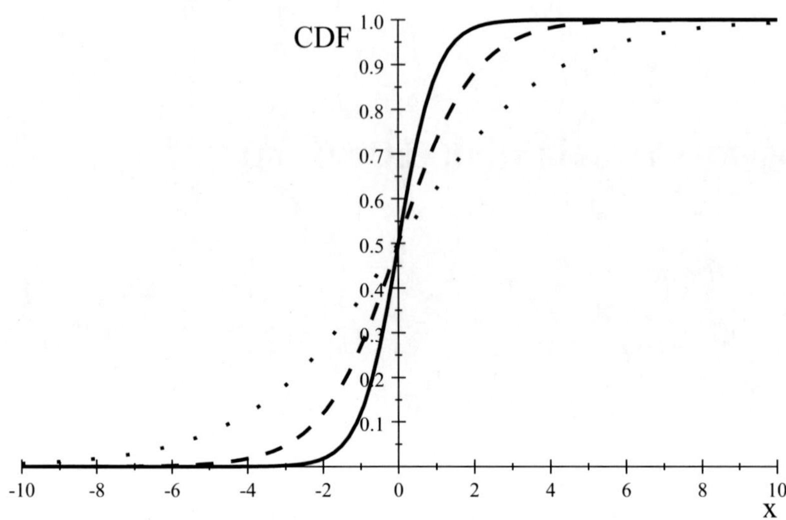

Figure 20.1.2. Cdfs - LO (0,0.5)-solid,, LO (0,1)-dash and LO (0,2) -dots.

Mean	μ
Median	μ
Variance	$\frac{\sigma^2\pi^2}{3}$
Coefficient of variance	$\frac{\pi\sigma}{\sqrt{3}\mu}$

For LO(0,1)

$E(X^{2m+1}) = 0.$ m= 0,1,2,…

$E(X^{2m}) = \pi^n(2^n - 1)|B_n|$

where B_n is the Bernoulli number

$$E(X^{2-}) = Var(X) = \frac{\pi^2}{3}$$

$$E(X^4) = \frac{3\pi^4}{15}$$

20.2. Basic Properties

Characteristic function, $\phi(t)$

$$\phi(t) = \int_{-\infty}^{\infty} e^{itx} \frac{e^{-x}}{(1+e^{-x})^2}.dx$$

Let u = $(1+e^{-x})^{-1}$then

$$\varphi(t) = \int_0^1 e^{itln\frac{u}{1-u}} \text{du}$$
$$=\int_0^1 u^{it}(1-u)^{-it}\text{du}$$
$$= B(1+it,1=it)$$
$$=\Gamma(1-it)\Gamma(1+it)$$

using the Euler's representation of gamma function

$$\Gamma(z) = \lim_{n\to\infty} \frac{n!n^z}{z(z+1)(z+2)...(z+n)} = \frac{1}{z}\prod_{n=1}^{\infty} \frac{(1+\frac{1}{n})^z}{1+\frac{z}{n}}$$, we have

$$\Gamma(1-it)\Gamma(1+it) = \prod_{n=1}^{\infty} (1+\frac{t^2}{n})^{-1}.\cdot$$

Moment generating function $\Gamma(1+t)\Gamma(1-t)$

Characteristic function $\Gamma(1+it)\Gamma(1-it)$

20.3. Distributional Properties

If X is LO (0,1), then

$$X \underline{\underline{d}} \prod_{j=1}^{\infty} W_i/\text{j}$$

where $W_1, W_2, \ldots$ are independent and identically distributed as Laplace LA (0,1)..

If X is LO (μ, σ), *then* $kZ + m$ *is* $LO\ (k\mu, |k|\sigma)$

If E (0,1), *then* $\mu + \sigma \ln(e^x - 1)$ *is* $LO\ (\mu.\sigma)$

20.4. Random Number Generation

Generate a random number U from uniform distribution, U (0,1).

Set $X = \mu + \sigma * (\ln U - \ln(1 - U))$.

Then X is a logistic random variable, LO $(\mu, \tau\sigma)$.

Chapter 21

Lognormal Distribution (LN (μ,σ))

21.1. Introduction

Pdf $\dfrac{1}{x\sigma\sqrt{2\pi}}e^{-\frac{1}{2}(\frac{\ln x-\mu}{\sigma})^2}$,x>0, $-\infty < \mu < \infty, \sigma > 0$.

The pdfs of LN (0,1), LN (0,2) and LN (0,4) are given in figure 21.1.1.

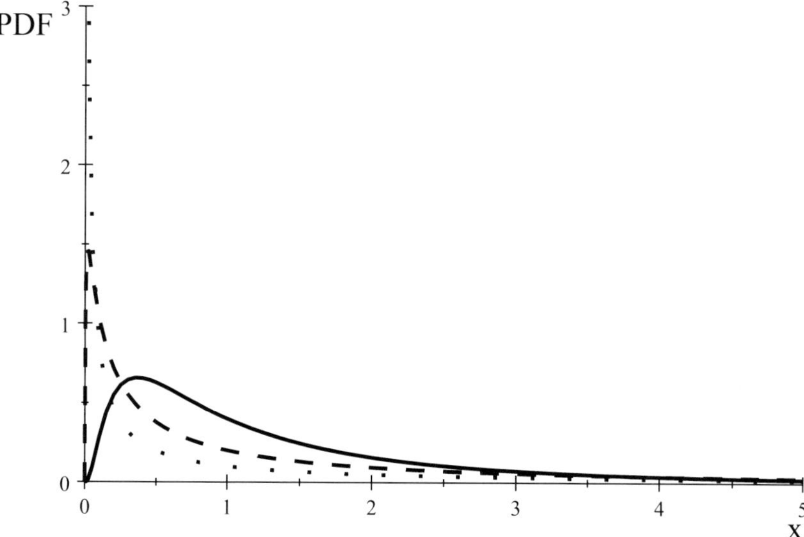

Figure 21.1.1. PDFs- LN (0,1)-solid, LN (0,2)-dash and LN (0,4)-dots.

CDF $\dfrac{1}{2}+\dfrac{1}{2}erf(\dfrac{\ln x-\mu}{\sigma\sqrt{2}})$

The CDFs of LN (0,1), LN (0,2) and LN (0,4) are given in figure 21.1.2.

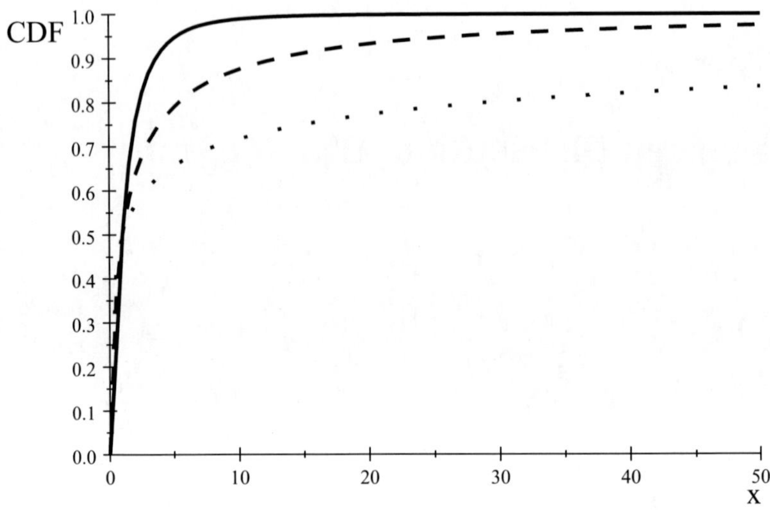

Figure 21.1.2. LN (0,1)-solid, LN (0,2)-dash and LN (0,4)-dots.

Mean	$e^{\mu+\sigma^2/2}$
Median	e^{μ}
Variance	$(e^{\sigma^2}-1)e^{2\mu+\sigma^2}$
Coefficient of variation	$\sqrt{e^{\sigma^2}-1}$
Moment about the origin E(X^r)	$e^{r\mu+\frac{1}{2}r^2\sigma^2}$

The Lognormal distribution is not uniquely determined by its moments. There is a family of distributions with same moments as the Lognormal distribution.

21.2. Basic Properties

Moment generating function $\sum_{n=0}^{\infty}\frac{t^n}{n!}e^{n\mu+n^2\sigma^2/2}$

Characteristic function $\sum_{n=0}^{\infty} \frac{(it)^n}{n!} e^{n\mu + n^2\sigma^2/2}$

The hazard rate function $\sqrt{\frac{2}{\pi}} \frac{e^{-(\frac{\ln x - \mu}{\sigma})^2}}{erfc(\frac{\ln x - \mu}{\sigma})}$

21.3. Distributional Properties

If X_1 *and* X_2 are independent LN $(\mu_i, \sigma_i{}_i^2)\ i = 1,2,.\ then$
X_1X_2 is LN $(\mu_1 + \mu_2, \sigma_1^2 + \sigma_2^2)$.
X_1/X_2 is LN $(\mu_1 - \mu_2, \sigma_1^2 + \sigma_2^2)$.
If X is LN (μ, σ^2), *then* aX *is* LN $(\mu + \ln a, \sigma^2)$.
If X is LN (μ, σ^2), *then* $\frac{1}{X}$ *is* $LN(-\mu, \sigma^2)$,
If X is LN (μ, σ^2), *then* X^α in LN$(a\mu, a^2\sigma^2)$
If X_1, X_2 ,...$X_n$are independent LN $(\mu_i, \sigma_i{}_i^2)\ i = 1,2, \ldots, n\ then$
$\Pi_{i=1}^n c_i X_i$ is LN $(\sum_{i=1}^n \mu_i, \sum_{i=1}^n \sigma_i^2)$
If X is distributed as N (μ,σ)), then e^X is distributed as LN (μ,σ).
If X is distributed as LN (μ,σ), then ln(X) is distributed as N (μ,σ).
If X_j is distributed as LN (μ,σ),j=1,2,...,n and X_j's are independent,

then Y = $(X_1X_2...X_n)$ is distributed as LN (μ, σ).),

21.4. Random Number Generation

Generate a normal random variable Z and *for a given* $\mu, \sigma,\ \infty < \mu < \infty$, $\sigma > 0$.

set $Y = \mu + \sigma * Z$
Return x -exp(Y).
Then X is a random variable from LN $(\mu. \sigma)$.

Chapter 22

Negative Binomial Distribution (NB(m,p))

22.1. Introduction

PMF $\binom{m+x-1}{m-1}\, q^m p^x$, x=0,1,2,…, 0<p <1, q= 1-p, m is an integer ≥ 1

The pmfs of NB (3,0.5) and NB (4,0.3) are given in figures 22.1.1 and 22.1.2.

Figure 22.1.1. PMF- NB (3,0.5).

CDF $\sum_{k=0}^{x}\binom{m+k-1}{m-1} q^m\, p^k$

The cdfs of NB ((3,0.5) and NB ((3,0.3) are given in figures 22.1.3. and 22.1.4.

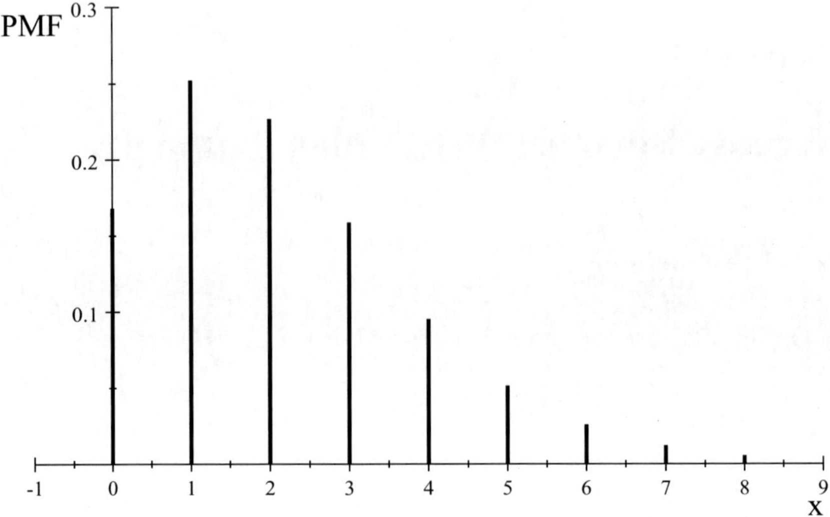

Figure 22.1.2. PMF- NB (5,0.3).

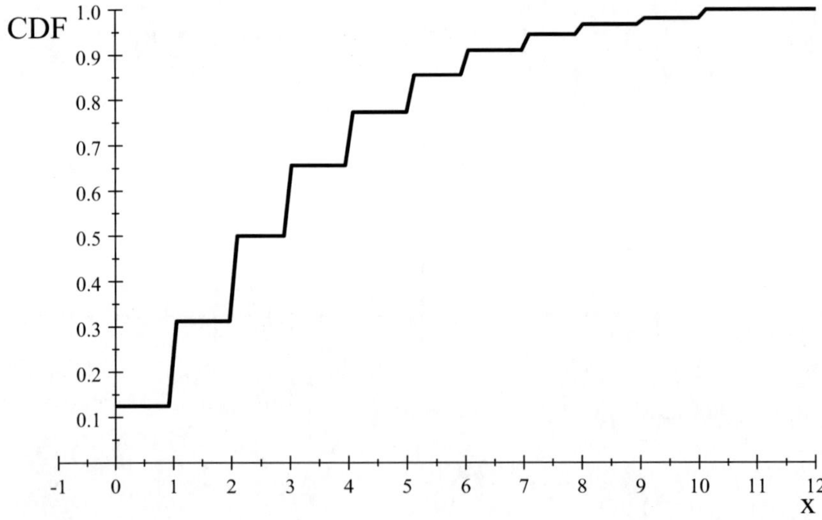

Figure 22.1.3. CGF- NB ((3,0.5).

Mean $\qquad \dfrac{mp}{1-p}$

Moments about the origin

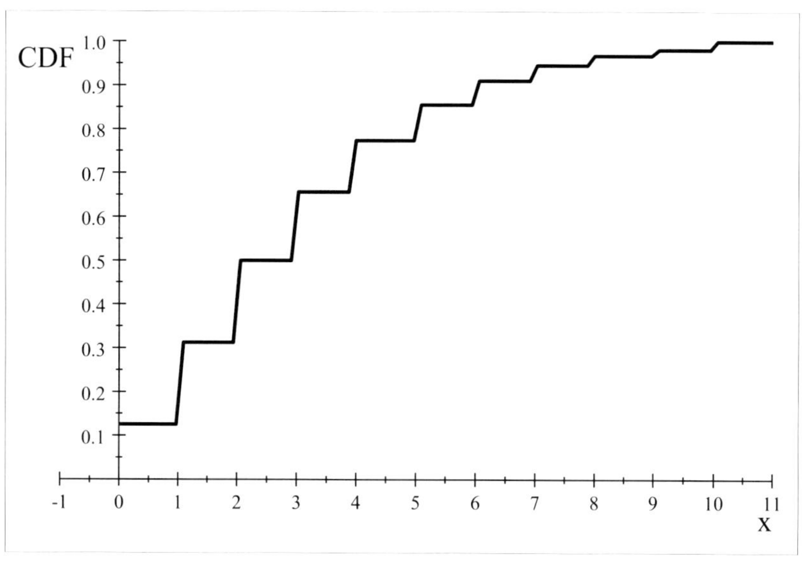

Figure 22.1.4. CDF- NB (5,0.3).

E(X^2) $\frac{mp(1+mp)}{(1-p)^2}$

E(X^{3-}) $\frac{mp[1+(3m\_1)+m^2p^2)]}{(1-p)^3}$

E(X^4)$\frac{mp[(1+(7m+4)p+(6m^2+4m+1)p^2+m^3p^3]}{(1-p)^4}$

Variance $\frac{mp}{(1-p)^2}$

Coefficient of variation $\frac{1}{\sqrt{mp}}$

22.2. Basic Properties

Moment generating function $(\frac{1-p}{1-pe^t})^m, t < -\ln p,$

Characteristic function $(\frac{1-p}{1-pe^{it}})^m$

22.3. Distributional Properties

If X_1 and X_2 are independent and are distributed as NB (m_1, p) and NB (m_2.P), then

X_1+X_2 is distributed as NB (m_1+m_2, p).

Let $\lambda = E(X) = \frac{mp}{1-p}$, then

$$P(X=k) = \frac{\lambda^k}{\mathrm{k!}} \frac{\Gamma(m+k)}{\Gamma(m)(m+\lambda)^k} \frac{1}{(1+\frac{k}{m})^m}$$

$$\lim_{m\to\infty} P(X=k) = \frac{\lambda^k}{\mathrm{k!}} \frac{\Gamma(m+k)}{\Gamma(m)(m+\lambda)^k} \frac{1}{(1+\frac{k}{m})^m} = \frac{\lambda^k}{\mathrm{k!}} e^{-\lambda}$$

Thus if m → ∞ and $\lambda = \frac{mp}{1-p}$, then NB (m,p) converges to Poisson distribution, P(λ).

NB(1,p) is GE(p).

If X_n is NB (m,p) and W_n

$$= \frac{X_n - \frac{mp}{1-p}}{\frac{\sqrt{mp}}{1-p}}$$,then W_n converges to normal,N(0,1).

22.4. Random Number Generation

Generate a sequence of random numbers from a uniform distribution.
Select a p (0<p1).
Count the numbers greater than and less than p.
When the number less than p first reaches m (a selected number), the number greater than p is a negative binomial, NB (m,p).

Chapter 23

Normal Distribution (N (μ,σ))

23.1. Introduction

The pdf $f_{\mu,\sigma}(x)$ $\frac{1}{\sigma\sqrt{2\pi}}e^{-\frac{1}{2}(\frac{x-\mu}{\sigma})^2}$ $, -\infty < \mu < x < \infty, \sigma > 0$

The graphs of pdfs of N (0,0.5) ,N (0,1) and N (0,3) are given in figure 23.1.

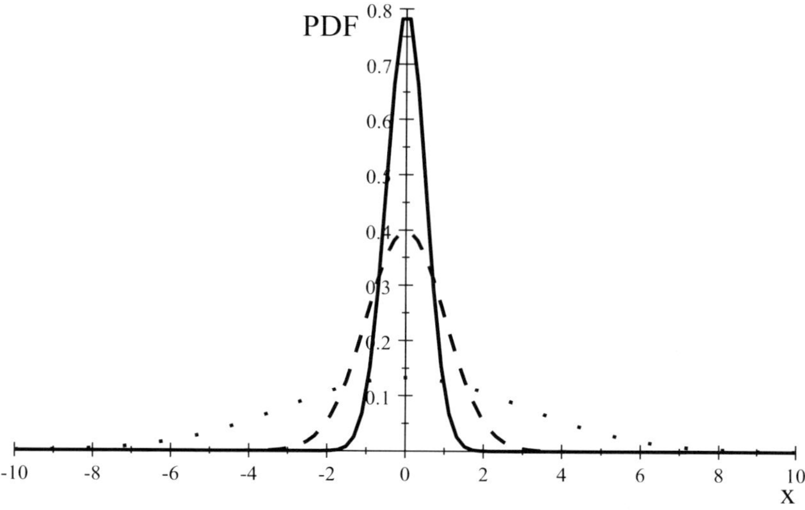

Figure 23.1. pdfs – N (0,0.5) -solid, N (0,1) -dash sand N (0,3) -dots.

CDF $F_{\mu,\sigma}(x)$ $\frac{1}{2}(1+\text{erf}(\frac{x-\mu}{\sigma\sqrt{2}}))$

The cdfs of N (0,0.5), N (0,1) and N (0,3) are given in figure 23.2.

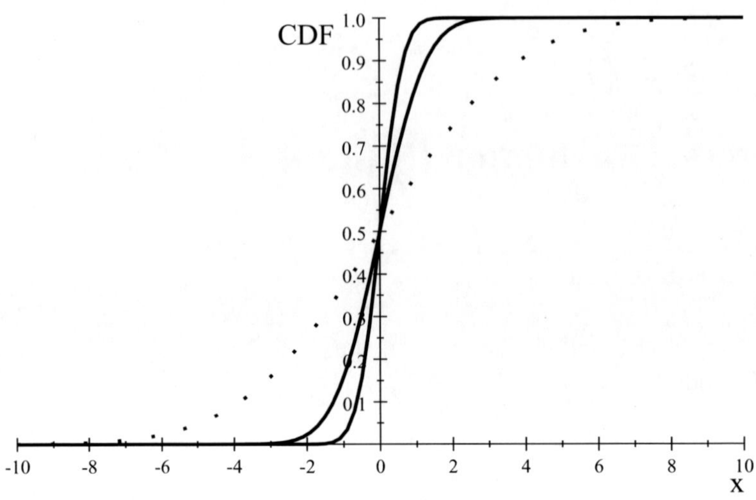

Figure 23.2. Cdfs – N (0,0.5)-solid, , N (0,1)-dash and N (0,3)-dots.

Mean	μ
Median	μ
Variance	σ^2
Entropy	$\ln(2\pi e\sigma^2)$
Coefficient of variation	$\frac{\sigma}{\mu}$

23.2. Basic Properties

MGF	$e^{\mu t+\frac{\sigma^2 t^2}{2}}$
CF	$e^{i\mu t-\frac{\sigma^2 t^2}{2}}$
Inflection points	$\mu \pm \sigma$

$$F_{\mu,\sigma}(\mu+n\sigma) - F_{\mu,\sigma}(\mu-n\sigma) = \text{erf}(\frac{n}{\sqrt{2}})$$

The following table gives values of $F_{\mu,\sigma}(\mu+n\sigma) - F_{\mu,\sigma}(\mu-n\sigma)$

for n=1 to 6

Table of $\text{erf}(\frac{n}{\sqrt{2}})$

n	$\text{erf}(\frac{n}{\sqrt{2}})$
1	*0.682689492*
2	*0.934499736*
3	*0.997300204*
4	*0.999937516*
5	*0.999999527*
6	*0.999999998*

$$(\frac{1}{x}-\frac{1}{x^2})\frac{1}{\sqrt{2\pi}}e^{-\frac{1}{2}x^2} \le \Phi(x) \le \frac{1}{\sqrt{2\pi}}e^{-\frac{1}{2}x^2}.$$

$$(\frac{1}{x}-\frac{1}{x^2})\frac{1}{\sqrt{2\pi}}e^{-\frac{1}{2}x^2} \le \Phi(x) \le \frac{1}{\sqrt{2\pi}}e^{-\frac{1}{2}x^2}.$$

Abramovitz and Segun's approximation for $\Phi(x) = F_{0,1}(x)$.

$$\Phi(x) = 1-\frac{1}{\sqrt{2\pi}}e^{-\frac{1}{2}x^2}(b_1t+b_2t^2+b_3t^3+b_4t^4+b_5t^5)+e(x),$$

$$t=\frac{1}{1+02326419\ x}, \ |e(x)| < 7.5\times10^{-8}$$

b_1*=0.31981539* b_2*= 0.356563782* b_3 *=1.781477937* b_4 *= 1.822155978*

b_5*= 1.330274420*

Moment about the origin

$E(X^k) =$

$$\sigma^k \sum_{k=1}^{(k+1)/2} \frac{k!\,\mu^{2k-1}}{(2i-1)!((k+1)/2-i)!2^{(k+1)/2-i`}\sigma^{2i-1}}$$

$$k = 1,3,5,...$$

$$\sigma^k \sum_{k=0}^{k/2} \frac{k!\,\mu^{2k-1}}{(2i)!(k/2-i)!2^{k/2-i`}\sigma^{2i}}$$

$$k = 2,4,6,...$$

Moment about the Mean

$$E(X-\mu)^k =$$

$$0, \quad k = 1.3,5,....$$

$$\sigma^k\,(k-1)!!, k = 2,4,6,....$$

where n!! denotes the double factorial.

Recurrence Relation of Moments

$$E(X^k) = (k-1)\sigma^2 E(X^{k-2}) + \mu E(X^{k-1})$$

$$k = 3,4,...$$

The statistic $\frac{\overline{X}_n - \mu}{S_n/\sqrt{n}}$ is distribute d as t - distributi on with n - 1 degees of freedom.

here $S_n = \sqrt{S_n^2}$.

3.3. Distributional Properties

Central Limit Theorem

Let X_1, X_2,,X_n be n independent and identically distributed random variables with finite mean μ and variance σ^2.

Define

$$\overline{X}_n = \frac{1}{n}(X_1 + X_2 + ... + X_n),\ then\ \frac{\overline{X}_n - \mu}{\sigma/\sqrt{n}} \to N(0,1)\ as\ n \to \infty.$$

Assume X_1, X_2, ...,X_n as a random sample from $N(\mu,\sigma)$.

Define

$$\overline{X}_n - \frac{1}{n}(X_1 + X_2 + \ldots + X_n) \text{ and } S_n^2 = \frac{1}{n-1}\sum_{i=1}^{n} (X_i - \overline{X}_n)^2.$$

It is known that $\overline{X}_n$and $S_n^2$are independent.

The converse is also true.

The binomial distribution BN(n,p)is approximated by

N (np/$\sqrt{(np(1-p))}$).

If $X_i$is $N(\mu_i, \sigma_i), i = 1,2,\ldots,n$,and X_i'sare independent, then *for any a_i, i=1,2,...,n*

$$\sum_{i=1}^{n} a_i X_i \text{ is } N\left(\sum_{i=1}^{n} a_i\mu_i, \sqrt{a_i^2\sigma_i^2}\right)$$

If X_1 and X_2 are independent continuous random variable such that X_1+X_2 and X_1-X_2 are independent, then X_1 and X_2 are normal.

If X and Z are independent random variable such that Z and X+Z are normal, then X is normal.

If X_1 and X_2 are independent and identically distributed random variables with finite means and for some $\alpha \neq 0,\pm 1$, and $E(X_1 - \alpha X_2 \mid \alpha X_1 + X_2) = 0$, then X_1 and X_2 are normal.

If X_1 and X_2 are independent and identically distributed random variables with finite second moments such that $E(X_1 - X_2 \mid X_1 + X_2) = 0$ and $E((X_1 - X_2)^2 \mid X_1 + X_2) = c$ where c is a constant, then X_1 and X_2 are normal,

If X_1, X_2 are independent random and identically distributed random variables such that

X_1 and $\frac{X_1 + X_2}{\sqrt{2}}$ have the same distribution, then X_1 is normal.

Let $X_{1,n} < X_{2,n} < \ldots < X_{n,n}$ be the order statistics of the n independent random variables $X_1, X_2 \ldots, X_n$ from N(0,1).then

$$E(X_{5,5}) = E(X_{1,5}) = \frac{15}{\pi\sqrt{\pi}} tan^{-1}(\sqrt{2}) - \frac{5}{2\sqrt{\pi}}$$

$$E(X_{n,n}) = E(X_{1,n}) = 1 + E(X_{1,n} X_{2,n})$$

23.4. Random Number Generation

Generate two independent random numbers U and V from uniform, UN (0,1) distribution.

Set $X = \sqrt{-2 \ln U}\, cos(2\pi V), Y = \sqrt{-2 \ln U}\, sin(2\pi V)$,

Then X and Y will be two independent N (0,1) random variables.

A better method

Let U_1 and U_2 be independent and uniformly distributed on (-1,1) with the condition

$$U_1^2 + U_2^2 \leq 1.$$

Set

$$X_1 = U_1\{\frac{-2\ln(U_1^2+U_2^2)}{U_1^2+U_2^2}\}^{1/2}$$

$$X_2 = U_2\{\frac{-2\ln(U_1^2+U_2^2)}{U_1^2+U_2^2}\}^{1/2}$$

Then X_1 and X_2 are independently distributed as N (0,1).
The following table gives the values of $\Phi(x)$ from 0 t0 3.0.

•ΘU

x	.00	.01	.02	.03	.04	.05	.06	.07	.08	.09
0	.5000	.5040	.5080	.5120	.5160	.5199	.5239	.5279	.5319	.5359
.1	.5398	.5438	.5478	.5517	.5557	.5596	.5636	.5675	.5714	.5753
.2	.5793	.5832	.5871	.5810	.5948	.5987	.6026	.6064	.6100	.6141
.3	.6179	.6217	.6255	.6293	.6331	.6369	.6406	.6443	.6480	.6517
.4	.6554	.6591	.6628	.6664	.6700	.6736	.6772	.6808	.6844	.6870
.5	.6915	.6959	.6985	.7019	.7054	.7088	.7123	.7157	.7190	.7224
.6	.7257	.7291	.7324	.7357	.7389	.7422	.7454	.7486	.7517	.7549
.7	.7580	.7611	.7642	.7673	.7704	.7734	.7764	.7794	.7823	.7852
.8	.7881	.7910	.7939	.7967	.7993	.8023	.8051	.8078	.8106	.8133
.9	.8159	.8186	.8212	.8238	.8264	.8259	.8315	.8340	.8365	.8389
1.0	.8413	.8438	.8461	.8485	.8508	.8531	.8554	.8577	.8599	.8621
1.1	.8643	.8665	.8686	.8708	.8729	.8749	.8770	.8790	.8810	.8830
1.2	.8849	.8869	.8888	.8907	.8925	.8944	.8962	.8980	.8997	.9015
1.3	.9032	.9049	.9066	.9082	.9099	.9115	.9131	.9147	.9162	.9177
1.4	.9192	.9207	.9222	.9236	.9251	.9265	.9279	.9292	.9306	.9319
1.5	.0332	.9345	.9357	.9370	.9382	.9394	.9406	.9418	.9429	.9441
1.6	.9452	.9463	.9474	.9484	.9495	.9505	.9515	.9525	.9535	.9545
1.7	.9554	.9564	.9573	.9582	.9591	.9599	.9608	.9616	.9625	.9633
1.8	.9641	.9649	.9656	.9664	.9671	.9678	.9686	.9693	.9899	.9706
1.9	.9713	.0719	.9726	.9732	.9738	.9744	.9750	.9756	.9761	.9767
2.0	.9772	.9778	.9783	.9788	.9793	.9798	.9803	.9808	.9812	.9817
2.1	.9821	.9826	.9830	.9834	.9838	.9842	.9846	.9850	.9854	.9857
2.2	.9861	.9864	.9868	.9871	.9875	.9878	.9881	.9884	.9887	.9890
2.3	.9893	.9896	.9898	.9991	.9904	.9906	.9909	.9911	.9913	.9915
2,4	.9918	.9920	,9922	.9925	.9927	.9929	.9931	.9932	.9934	.9936
2.5	.9938	.9940	.9941	.9943	.9945	.8846	.9948	.9949	.9951	.9952
2.6	.9953	.9955	.9956	.9957	.9959	.9960	.9961	.9962	.9963	.9964
2.7	.9965	.9966	.9967	.9968	.9969	.9970	.9971	.9972	.9973	.9974
2.8	.9974	.9975	.9976	.9977	.9977	.9978	.9979	.9979	.9980	.9981
2.9	.9981	.9982	.9982	.9983	.9984	.9984	.9985	.9985	.9986	.9986
3.0	.9986	.9987	.9987	.9988	.9988	.9989	.9989	.9989	.9990	.9990

Chapter 24

Pareto Distribution (PA (σ, δ))

24.1. Introduction

Pdf $\qquad \dfrac{\delta\sigma^{\delta}}{x^{\delta+1}}, x > \sigma, \delta > 0.$

The pdfs of PA (1,1/2), PA (1,1) and PA (1,3) .are given in figures 24.1.1.

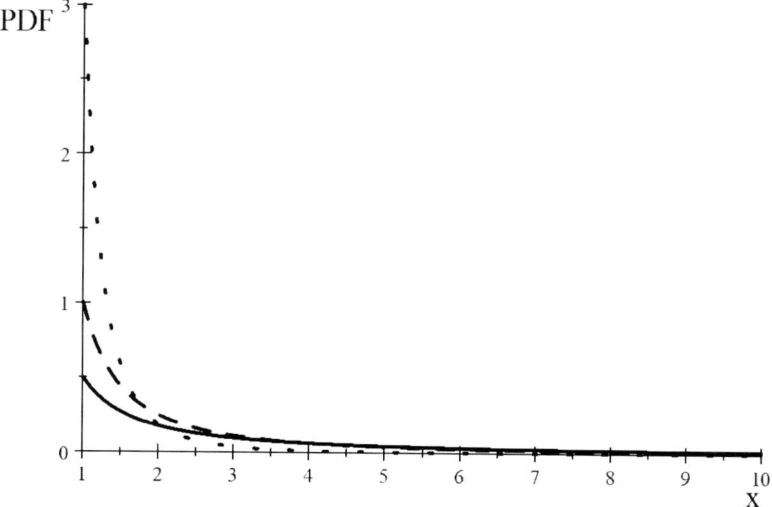

Figure 24.1.1. PDFS- PA (1,1/2)- solid, PA (1,.1) -dash and PA (1,.3)-dots.

Cdf $\qquad 1-(\dfrac{\sigma}{x})^{\delta}$

The cdfs of PA (1,1/2) and PA (1,1) and PA (1,3) are given in figure 24.1.2.

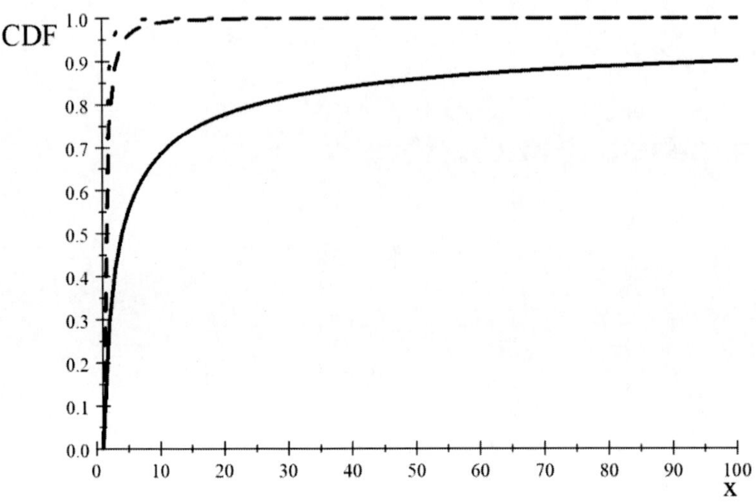

Figure 24.1.2. Cdfs - PA (1,1/2)- solid, PA (1,1)-dash and PA (1,3) -dots.

Mean $\frac{\sigma\delta}{\delta-1}, \delta>1$

Variance $\frac{\sigma^2\delta}{(\delta-1)^2(\delta-2)}, \delta>2$

Median $\sigma\, 2^{1/\delta}$

Coefficien t of variation $\sqrt{\frac{1}{\delta(\delta-2)}}, \delta>2$

24.2. Basic Properties

Moment about the origin $E(X^k)=\frac{\delta\sigma^k}{\delta-k}, \delta>k$

Moment generating function does not exist

Entropy $1+\frac{1}{\delta}-\ln\delta-(1+2\delta)\ln\sigma$

24.3. Distributional Properties

If $X_1, X_{2,\dots,}X_n$ are independent PA (σ,δ) , *then*

$2\delta ln\ (\frac{\Pi_{I=1}^{n}X_i}{\sigma^n}$ is distributed as chi-square with 2n degrees of freedom.

Different Types of Pareto Distribution
Pareto Type i

$$F(x)=1-(\frac{x}{\sigma})^{-\alpha}, x>\sigma, \alpha>0$$

Pareto Type II

$$F(x)=1-(1+(\frac{x-\mu}{\sigma}))^{-\alpha}, x>\mu, \sigma>0, \alpha>0$$

Pareto Type III

$$F(x)=1-(1+(\frac{x-\mu}{\sigma})^{1/\gamma})^{-1}, x>\mu, \sigma>0, \gamma>0$$

Pareto Type iv

$$F(x)=1-(1+(\frac{x-\mu}{\sigma})^{1/\gamma})^{-\alpha}, x>\mu, \sigma>0, \gamma>0, \alpha>0.$$

Generalized Pareto distribution PA(σ,δ,μ)

Pdf $f_{\alpha,\sigma,\delta}$ $\frac{1}{\sigma}(1+\delta(\frac{x-\mu}{\sigma})^{-(1+1/\delta)}, \delta \neq 0,$

$x \geq \mu, \text{for } \delta > 0, \mu < x < \mu - \sigma/\delta, \text{for } \delta < 0$

$\frac{1}{\sigma} e^{-(\frac{x-\mu}{\sigma})}$, $x > \mu$ for $\delta = 0$

If $\delta = -1$, then PA$(\sigma, -1, \mu)$ is a uniform distributi on, UN$(\mu, \mu + \sigma)$.

24.4. Random Number Generation

Generate a uniform random variable from uniform, UN (0,1).

Set $X = \delta(1 - U)^{-1/\sigma}$

PA $(0,\sigma)$

Chapter 25

Poisson Distribution (P(λ))

25.1. Introduction

PMF $$\frac{\lambda^x e^{-\lambda}}{x!}, x = 0.1.2..., \lambda > 0.$$

The PMF of P (0.6) and P (3) are given in figure 25.1.1. and 25.1.2.

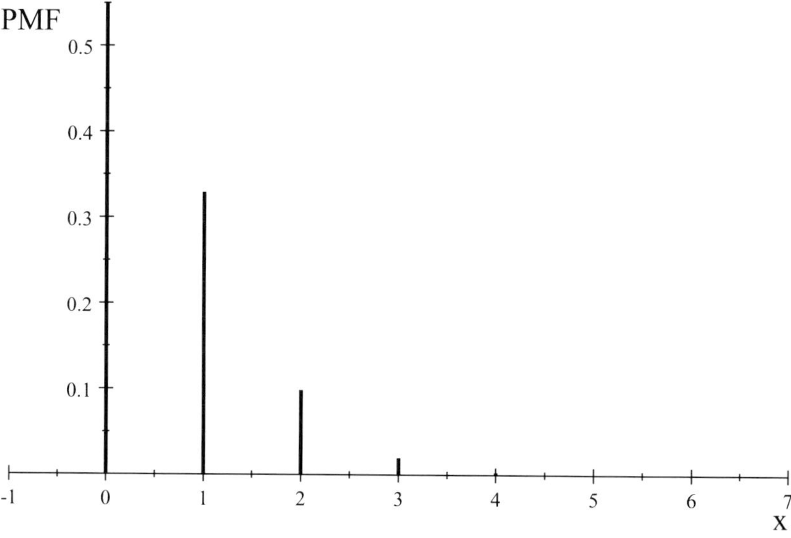

Figure 25.1.1. PMF- P (0.0.3).

Cdf $$\sum_{k=0}^{x} \frac{e^{-\lambda}\lambda^k}{k!}$$

The cdfs of P (0.6) and P (2) are given in figures 25.1.3 and 25.1.4.

Figure 25.1.2. PMF- P(3).

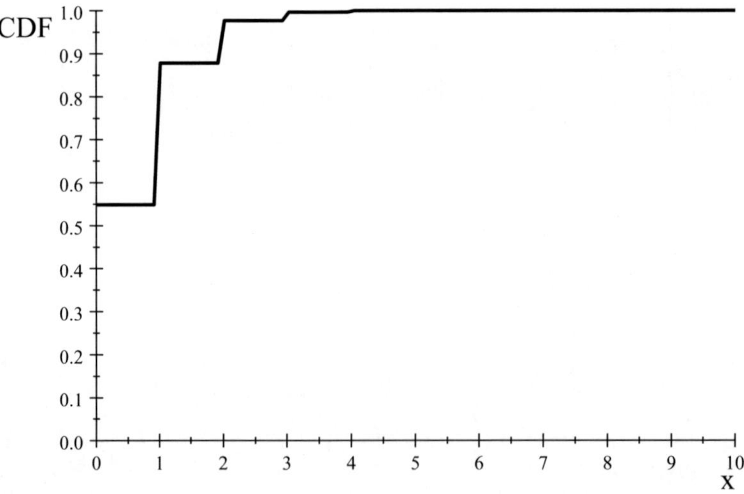

Figure 25.1.3. CDF-P (0.6).

Mean	λ
Variance	λ
Coefficient of variation	$\frac{1}{\sqrt{\lambda}}$
kurtosis	$3+\frac{1}{\lambda}$
Factorial moments	$E(\mathrm{H}_{i=1}^{k}(x-i+1))\ \lambda^k$
Moments about mean	$\mu_k = E(X-\lambda)^k$

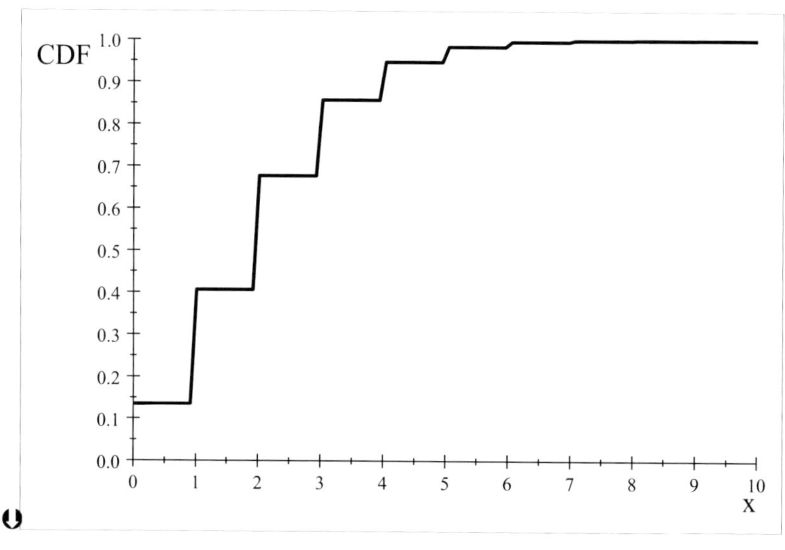

Figure 25.1.4. CDF-P (2).

$$\mu_1 = 0$$

$$\mu_2 = \lambda$$

$$\mu_3 = \lambda$$

$$\mu_4 = \lambda + 3\lambda^2$$

$$\mu_{r+1} = r\lambda\mu_{r-1} + \lambda \frac{d\mu_r}{d\lambda}, r = 2,3,...$$

$$\mu_r = \lambda \sum_{k=0}^{r-1} \binom{r-1}{k} \mu_k, \; r = 2,3,...$$

Coefficient of variation $\qquad \frac{1}{\sqrt{\lambda}}$

25.2. Basic Properties

Moment generating function $\qquad e^{\lambda(e^t - 1)}$

Characteristic function $e^{\lambda(e^{it}-1)}$

Probability generating function $e^{\lambda(t-1)}$

Let X_1 *is* $P(\lambda_1)$ *and* X_2 is $P(\lambda_2)$.
We assume X_1*and*X_2 *as independent.*

$$P(X_1 = m\}|X_1 + X_2{=}n)$$
$$= \frac{P(X_1{=}m,, X_1{+}X_2{=}n)}{P(X_1{+}X_2{=}n)}$$
$$= \frac{P(X_1{=}m,)(X_2{=}n{-}m)}{P(X_1{+}X_2{=}n)}$$
$$= \frac{e^{-\lambda_1`\lambda_1^m}}{m!} \frac{\lambda_2^{n-m}}{(n-m)!} \frac{n!}{e^{-(\lambda_1+\lambda_2)}} \frac{1}{(\lambda_1+\lambda_2)^n}$$
$$= \binom{n}{m} \left(\frac{\lambda_1}{\lambda_1+\Gamma_2}\right)^m \left(\frac{\lambda_2}{\lambda_1+\Gamma_2}\right)^{n-m}$$

Therefore, the conditional distribution of X_1 *given* $X_1 + X_2$ is Binomial,

BN (n, $\frac{\lambda_1}{\lambda_1+\Gamma_2}$)

25.3. Distributional Properties

$$P(X \geq k) = \int_0^\lambda \frac{e^{-u} u^{k-1}}{(k-1)!} du, k = 1,2,\ldots$$

$$\frac{P(X=k+1)}{P(X=k)} = \frac{\lambda}{x+1}, \text{ k=0,1,2,...}$$

Let X be distributed as B(n,p),then the PMF of X is

$$p(x{=}k) = \frac{n!}{k!(n-k)!} p^k (1-p)^{n-k}.$$

Using the relation

$$n! \cong \sqrt{2\pi n}\, n^n e^{-n} \text{ for large n,}$$

and

$$\lim_{n\to\infty}(1-\frac{k}{n})^{n}=e^{-k}.$$

We have for μ=np

$$lim_{n\infty}\frac{n!}{k!\,(n-k)!}p^{k}(1-p)^{n-k}$$

$$\lim_{n\to\infty}=\frac{1}{k!}\sqrt{\frac{n}{n-k}}\frac{e^{-k}}{(1-\frac{k}{n})^{n}}\mu^{k}(1-\frac{\mu}{n})^{n-k}$$

$$\rightleftarrows$$

$$=\frac{\mu^{k}e^{-\mu}}{k!}$$

Thus as n tends to infinity, the PMF of X tends to PMF of Poisson distribution with finite μ= np.

25.4. Random Number Generation

Generate $U_1, U_2, \ldots U_k$ random numbers from uniform distribution, UN (0,1) if

$$\prod_{i=1}^{k}U_{i}\leq e^{-\mu},\ \text{for some given } \mu,$$

then k is a random number from Poisson distribution, P (μ).

Chapter 26

Power Function Distribution (POW(a,b,δ))

26.1. Introduction

Pdf $\frac{\delta}{b-a}(\frac{x-a}{b-a})^{\delta-1}, -\infty < a \leq x \leq b, \delta > 0$

The pdfs of POW (0,1,3), POW (0,1,6) and POW (0,1,8) are given in figure 26.1.1.

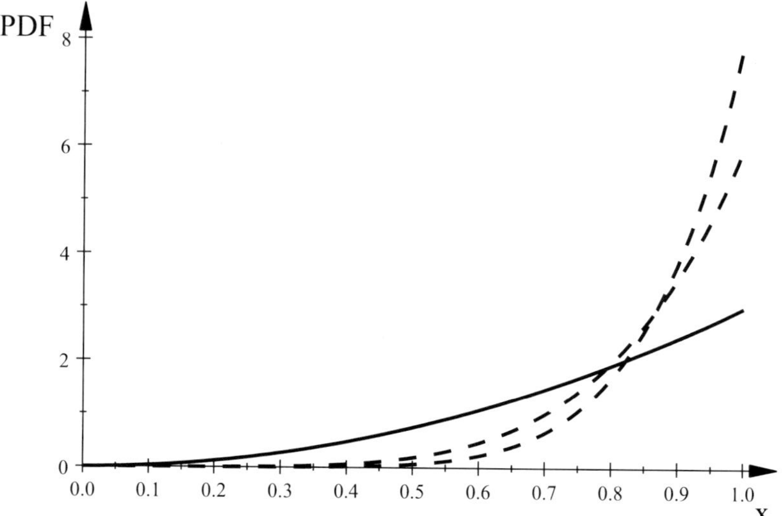

Figure 26.1.1. pdfs= POW (0,1,3)=solid,, POW (0,1,6)-dash and POW (0,1,8)-dots.

Cdf $(\frac{x-a}{b-a})^{\delta}$

The cdfs of POW (0,1,3), POW (0,1,6) and POW (0,1,8) are given in figure 26.1.2.

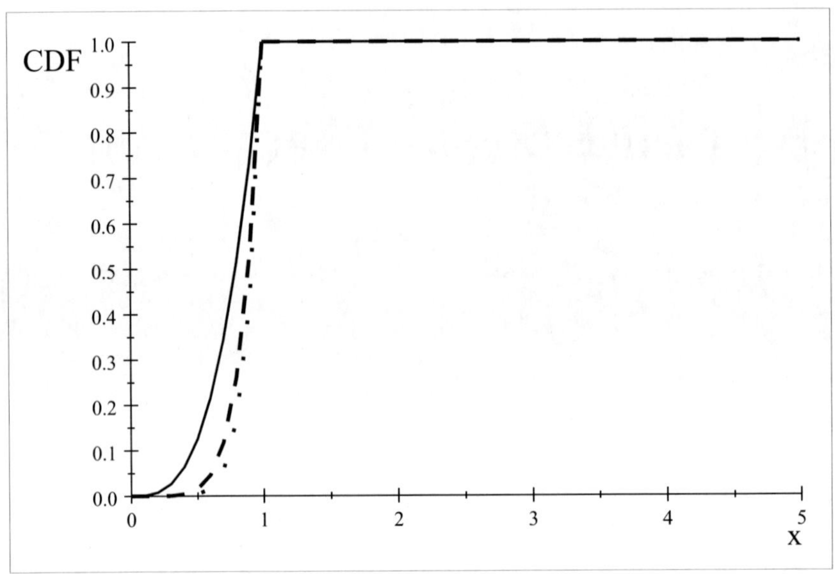

Figure 26.1.2. CDFs- POW (1,3)-solid, POW (1,6)-dash and POW (1,8)-dots.

26.2. Basic Properties

Mean	$a+\frac{\delta}{\delta+1}(b-a)$
Median	$a+(b-a)2^{-1/\delta}$
Variance	$\frac{\delta\,(b-a)^2}{(\delta+1)^2(\delta+2)}$
Entropy	$\frac{\delta\text{-}1}{\delta}-\ln\frac{\delta}{b-a}$
$E(X^r)$	$\sum_{i-0}^{r}\binom{r}{i}a^{r-i}\frac{\delta}{\delta+i}(b-a)^i$
Hazard rate function h(x)	$\frac{\delta(x-a)^{\delta}}{(b-a)^{\delta}-(x-a)^{\delta}}$

26.3. Distributional Properties

The moment generating function M(t) of POW (0,1,α) *is*

$$M(t) - \alpha \int_0^1 x^{\alpha-1} e^{tx}\, dx = \sum_{k=0}^{\infty} \int_0^1 \frac{(tx)^k x^{\alpha-1}}{k!}\, dx$$
$$= \sum_{k=0}^{\infty} \frac{t^k}{k!(k+\alpha)!}.$$

The characteristic function $\phi(t) is$

$$\phi(t) = \sum_{k=0}^{\infty} \frac{(it)^k}{k!(k+\alpha)!}$$

If ð=1, then POW(a,b,1) becomes uniform distribution, UN (a,b).
If a=1 and b=1, POW (a,b,1) coincides with beta distribution BE (1,ð)

26.4. Random Number Generation

Generate a uniform random number from UN (0,1).

Set X $= a + (b - a)U^{1/\delta}$.

Then X is a POW (a, b, δ) random variable.

Chapter 27

Student's t Distribution (ST(n))

27.1. Introduction

Pdf $\frac{1}{\sqrt{n}B(n/2,1/2))}(1+\frac{x^2}{n})^{-(n+1)/2}.-\infty < x < \infty,$

n is known as degrees of freedom,

The pdfs of ST (1), ST (3) and ST (25) are given in figures 27.1.1.

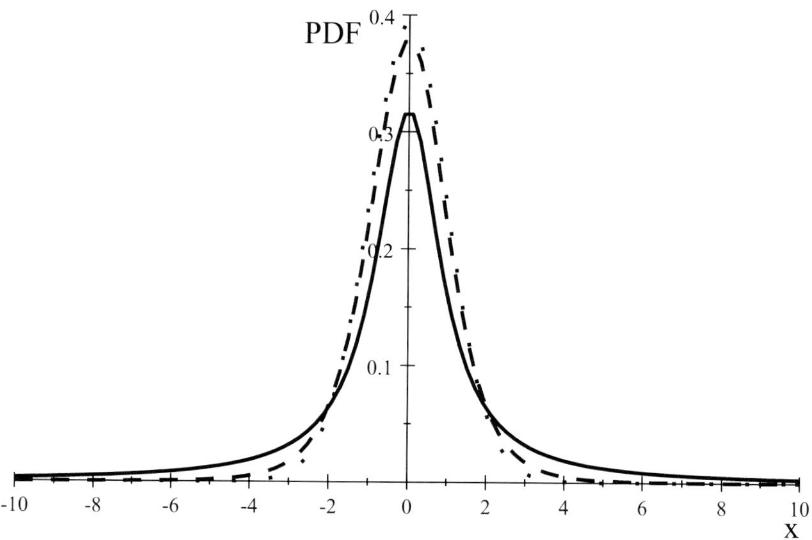

Figure 27.1.1. pdfs - ST (1)-solid, ST (5)-dash and ST (25)-dots.

The CDFs of ST (1), ST (3) and ST (25) are given in figure 27.2.

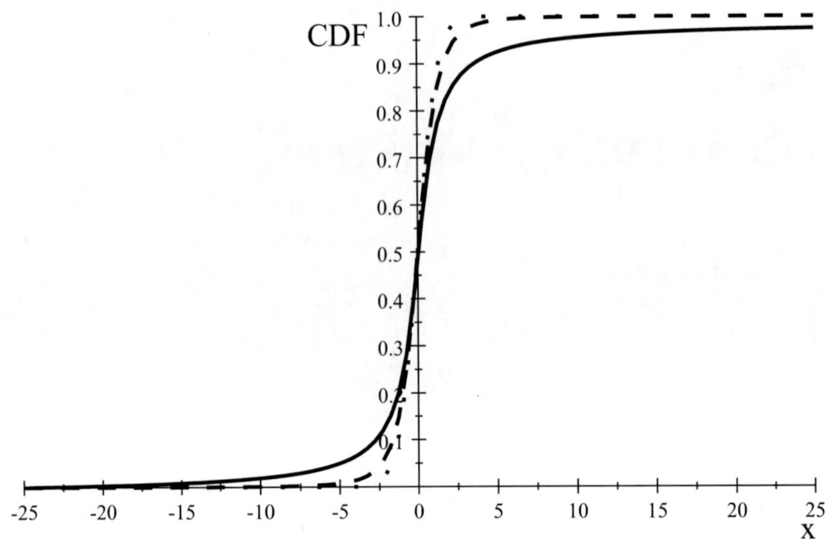

Figure 27.1.2. cdfs- ST (1)-solid, ST (3)=dash) and ST(25)-dots.

Pdf for ST (1)	$\frac{1}{\pi}(1+x^2)^{--1}.-\infty < x < \infty$
Pdf for ST (2)	$\frac{1}{(2+x^2)^{3/2}}$
Mean	0 if n>1, undefined for n=1
Variance	$\frac{n}{n-2}, n>2$
Median	0

The points of inflections $\pm\sqrt{n(n+2)}$

27.2. Basic Properties

Moment about the origin $E(X^k)$ =

$$0 \text{ if k is odd, } k < \text{n.}$$

$$\prod_{i=1}^{k/2} \frac{2i-1}{n-2i} n^{k/2}, \text{if k is even,} 0 < k < n$$

$$\frac{1}{2}+\frac{1}{\pi}\tan^{-1}\left(\frac{x}{\sqrt{n}}\right)+\frac{1}{\pi}\frac{x\sqrt{n}}{n+x^2}\sum_{j=0}^{(n-3)/2}\frac{a_j}{(1+\frac{x^2}{n})^j}, n \text{ odd}, n \geq 3$$

Pdf

$$\frac{1}{2}+\frac{1}{\pi}\tan^{-1}(x), n=1$$

$$\frac{1}{2}+\frac{x}{2(n+x^2)^{1/2}}\sum_{j=0}^{(n-2)/2}\frac{b_j}{(1+\frac{x^2}{n})^j}, n \text{ odd}$$

where

$$a_j = \frac{2^j j!}{(2j+1)(2j-1)\ldots 3}, b_j = \frac{(2j-1).(2j-3)\ldots.1}{2^j j!}$$

27.3. Distributional Properties

Characteristic function

$$\varphi_n(t) = \frac{2}{\pi\Gamma(n/2)}\left(\frac{|t|}{2\sqrt{n}}\right)^{n/2} Y_{n/2}\left(\frac{|t|}{\sqrt{n}}\right)$$

where $Y_{n/2}(.)$is the Bessel function of the second kind

In particular for odd n

$$\phi_n(t) = \sum_{j=0}^{m-1} c_{j,m} |t\sqrt{n}|^j e^{-|t\sqrt{n}|}$$

$c_{0,m} = 1$, $c_{1,m} = 1$, $c_{m-1,m} = ((2m-3)(2m-5)\ldots 3.1)^{-1}$

$c_{j,m} = (c_{j-1,m-1} + (2m-3-j)\, c_{j,m-1})(2m-3)^{-1}$, $m = (n+1)/2$

In particular

$$\varphi_1(t) = e^{-|t|}$$

$$\varphi_3(t) = (1+|t\sqrt{3}|)e^{-|t\sqrt{3}|}$$

$$\varphi_5(t) = (1+|t\sqrt{3}|+\frac{5}{2}t^2)e^{-|t\sqrt{5}|}$$

Let T_n be the ST (0,n) random variable then

$$P(|T_{2n+1}| \leq t) = \frac{2}{\pi}[\theta + \sin\theta)\cos\theta + \frac{2}{3}\cos^2(\theta) + \frac{2.4....2n-2}{1.3....2n-1}\cos^2\theta$$

$$P(|T_{2n}| \leq t) = \sin\theta\,\{1 + \frac{1}{2}\cos^2(\theta) + \frac{1,3}{2.4}\cos^4(\theta) + \frac{1,3,5....2n-3}{2.4.....2n=2}\cos^{2n-2}(\theta)$$

where $\theta = \tan^{-1}(t/\sqrt{n})$.

If Y is distributed as F (m,m), then $\frac{\sqrt{m}}{2}(Y^{1/2} - Y^{-1/2})$ is distributed as ST (m).

27.4. Random Number Generation

Generate a Standard normal, N (0,1) random variable Z

Generate a Gamma, GM (n/2.2) random variable Y

Set $X = \frac{Z}{\sqrt{Y/n}}$.

Then X is a ST (n) random variable.

Chapter 28

Rayleigh Distribution (RA (μ,σ))

28.1. Introduction

Pdf $\frac{x-\mu}{\sigma}\exp(-\frac{1}{2}(\frac{x-\mu}{\sigma})^2)$, $0\le \mu < x < \infty, \sigma > 0.$

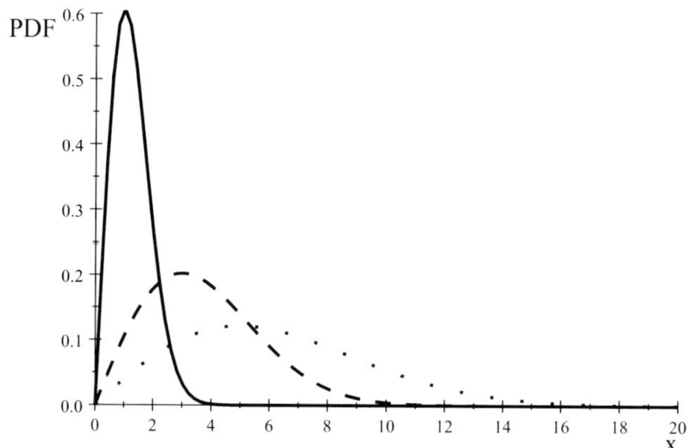

Figure 28.1.1. pdfs RA (0,1)-solid, RA (0,3)-dash and RA (0,5)-dots.

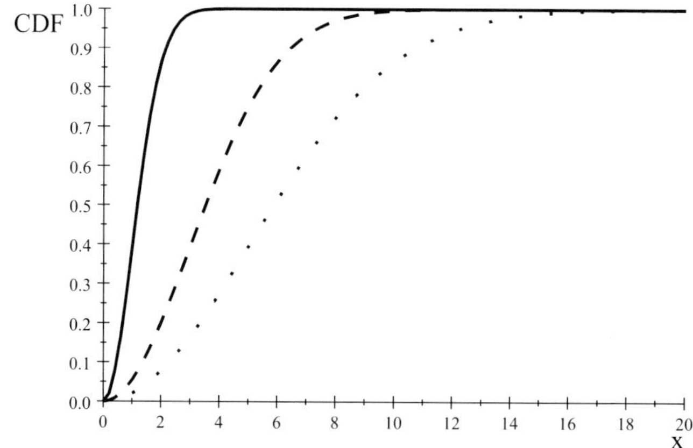

Figure 28.1.2. cdfs - RA (0,1)- solid, RA (0,3)=dash and RA (0,5) -dots.

The pdfs of RA (0,1), RA (0,3) and RA (0,5) are given in figure 28.1.1.

CDF $1\text{-}e^{-\frac{1}{2}(\frac{x-\mu}{\sigma})^2}$

The cdfs of RA (0,1), RA (0,3) and RA (0,5) are given in figure 28.1.2.

The percentile points of RA (0,1/2). RA (0,1) and RA (0,2) are given below.

Table 28.1. Percentiles RA (0,1/2).RA (0,1) and RA(0,2)

P	RA (0,1/2)	RA (0,1)	RA (0,2)
0.05	0.1502	0.3204	0.6406
0.1	0.2295	0.4590	0.9181
0.2	0.3340	0.6681	1.3361
0.3	0.4223	0.8446	1.6892
0.4	0.5054	1.0108	2.0215
0.5	0.5887	1.1774	2.3548
0.6	9.6769	1.3537	2.7075
0.7	0.7759	1.5518	3.1035
0.8	0.8971	1.7941	3.5883
0.9	1,0730	2.1460	4.2919
0.95	1.2239	2.4477	4.8955

Mean $\mu+\sigma\sqrt{\frac{\pi}{2}}$

Median $\mu+\sigma\sqrt{ln\,4}\approx\mu+1.1774\,\sigma$

Mode $\mu+\sigma$

Variance $\frac{4-\pi}{2}\sigma^2$

Entropy $1+\frac{\gamma}{2}+\ln(\frac{\sigma}{\sqrt{2}})$

Hazard Function $\frac{x-\mu}{\sigma^2}$

The hazard rate function increases with x.

28.2. Basic Properties

Moment about μ E(X-μ)^k $\sigma^k 2^{k/2} \Gamma(\frac{k+2}{2})$

MGT $e^{\mu t}(1\text{-}\sigma e^{\frac{\sigma^2 t^2}{2}} \sqrt{\frac{\pi}{2}} (\text{erf}\,(\frac{\sigma t}{\sqrt{2}} + 1))$

where erf is the error function.

CF $e^{i\mu t}(1\text{-}\sigma e^{\frac{-\sigma^2 t^2}{2}} \sqrt{\frac{\pi}{2}} (-i\text{erf}\left(\frac{i\sigma t}{\sqrt{2}}\right) - i))$

where erfc is the complementary error function.

28.3. Distributional Properties

If X_1 and X_2 are independent and are distributed as $N(0, \sigma)$, then

$Y = (X_1^2 + X_2^2)^{1/2}$is distributed as $RA\ (0, \sigma)$.

If X is distributed as $RA\ (\mu, \sigma)$, then $(x - \mu)^2$ is distributed $\text{E}(0\ ,2\sigma^2)$.
If X is distributed as RA (0,1), then X^2 is distributed as Chi (2).

28.4. Random Number Generation

Generate U from a uniform distribution on (0,1).

Set $X = \sigma\sqrt{-2\ln(1-U)}$, then X is RA (0,$\sigma$).

Chapter 29

Uniform Distribution (UN(*a,b*))

29.1. Introduction

Pdf $\frac{1}{b-a}, -\infty < a \le x \le b < \infty$

The pdf of UN (0,5) for a-0 and b=5 is given in figure 29.1.

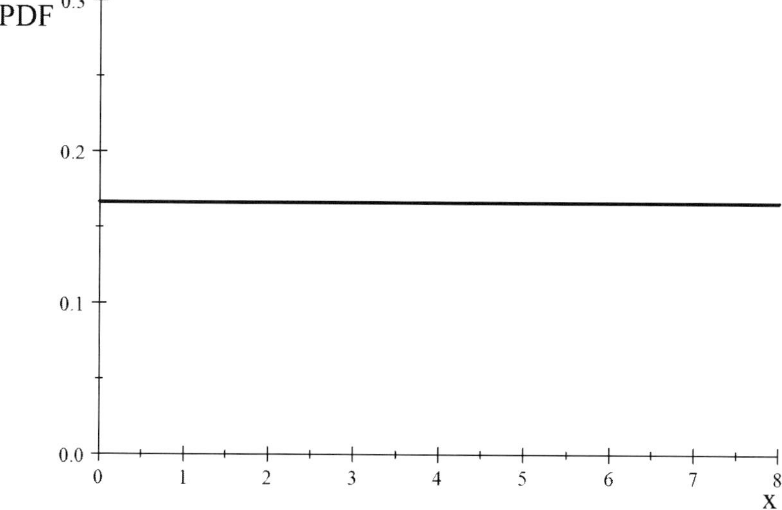

Figure 29.1. PDF - **UN (0,6).**

CDF $\frac{x-a}{b-a}, a \le x \le b$

$1, x \ge b$

The CDF of UN (0,6) is given in figure 29.2.

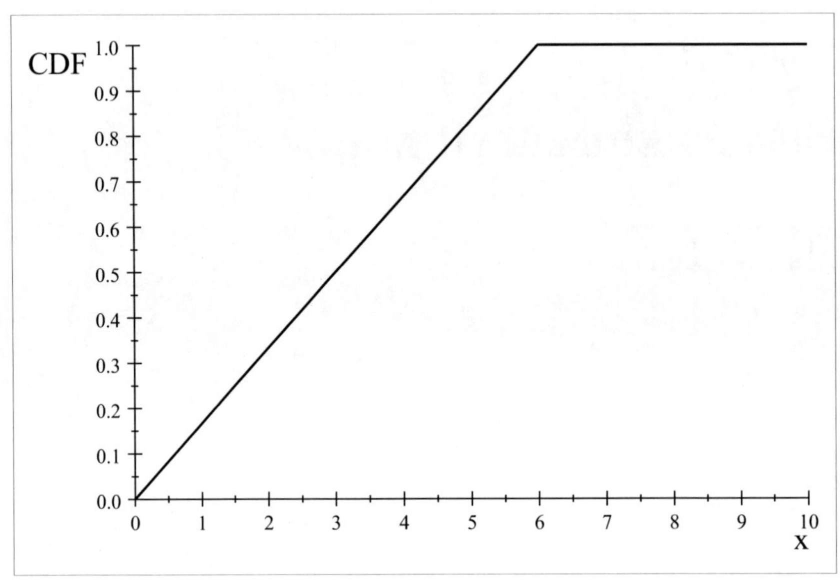

Figure 29.2. cdf- UN (0,6).

Mean	$\frac{a+b}{2}$
Median	$\frac{a+b}{2}$
Variance	$\frac{(b-a)^2}{12}$
Skewness	0
Entropy	ln(b-a)

29.2. Basic Properties

Moment generating function $((m_{a,b}(t)))$ $\frac{e^{bt}-e^{at}}{t(b-a)}$

$m_{0,b}(t)$ $\qquad e^{bt/2}\dfrac{\sinh(bt/2)}{bt/2}$

Characteristic function ($\phi_{a,b}(t)$) $\qquad \dfrac{e^{ibt}-e^{iat}}{it(b-a)}$

$\phi_{0,1}(t)$ $\qquad \sum_{j=0}^{\infty}\dfrac{(it)^j}{(j+1)!}$

$\phi_{-1,1}(t)$ $\qquad \dfrac{\sin t}{t}$

n-th moment about 0

$E(X^n)$ $\qquad \dfrac{1}{n+1}\sum_{i=0}^{n} a^i b^{n-i}$

The hazard rate function

$\dfrac{1}{b-x}$

29.3. Distributional Properties

Suppose X_1 and X_2 are both independently distributed as UN (0,1) and $Y = X_1+X_2$, then the pdf f(y) of Y is given by

f(y) $\qquad$ y, $0<y\leq 1$

$\qquad$ 2-y, $1\leq y\leq 2$

0.otherwise.

If X has the uniform distribution, UN (0,1), then -2ln(X) has the chi-square distribution with 2 degrees of freedom.

If $X_1, X_2,\ldots,X_n$ are independent and identically distributed as UN (0,1), then the kth order statistics, $X_{k,n}$ is distributed as Beta, BE (k,n-k+1).

If $X_1, X_2, \ldots, X_n$ are independent and identically distributed as UN (0,1) and $Y_n = X_1+X_2+\ldots+X_n$ then the pdf of Y_n is

$$f_Y(y) = \frac{1}{(n-1)!}\sum_{j=0}^{k}(-1)^j \binom{n}{j} (y-j)^{n-1}$$

for $k \le y \le k+1$, k=0,1,,…,n-1 = 0, otherwise.

29.4. Random Number Generation

Generate a uniform random number U, from UN (0,1).

Select a and b, $-\infty < a < b < \infty$.
Put x = a+(b-a) U. then X is a random variable from UN (a,b).

Chapter 30

Weibull Distribution (WE(μ, σ, δ))

30.1. Introduction

Pdf $\qquad \frac{\delta}{\sigma}(\frac{x-\mu}{\sigma})^{\delta-1} e^{-(\frac{x-\mu}{\sigma})^{\delta}}, -\infty < \mu < x < \infty, \sigma, \delta > 0.$

The pdf decreases for ð≤1 and is unimodal for ð>1.

WE ($\mu,\sigma,1$)is the exponential distribution, E (μ, σ).

The pdfs of WE (0,1,2), WE (0,1,4) and WE (0,1,6) are given in figure 30.1.1.

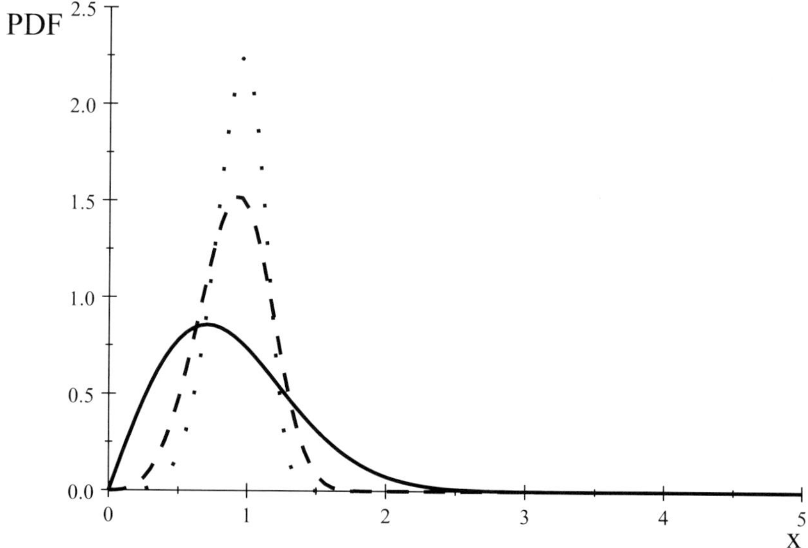

Figure 30.1.1. pdfs- WE (0,1,2)-sold, WE (0,1,4)-dash and WE (0,1,6)-dots.

Cdf $F_{(\mu,\sigma,\delta\}}(x)$ $\quad 1-e^{-(\frac{x-\mu}{\sigma})^{\delta}}$

The CDFs of WE (0,1,2), WE (0,1,4) and WE (0,1,6) are given in figure 30.1.2.

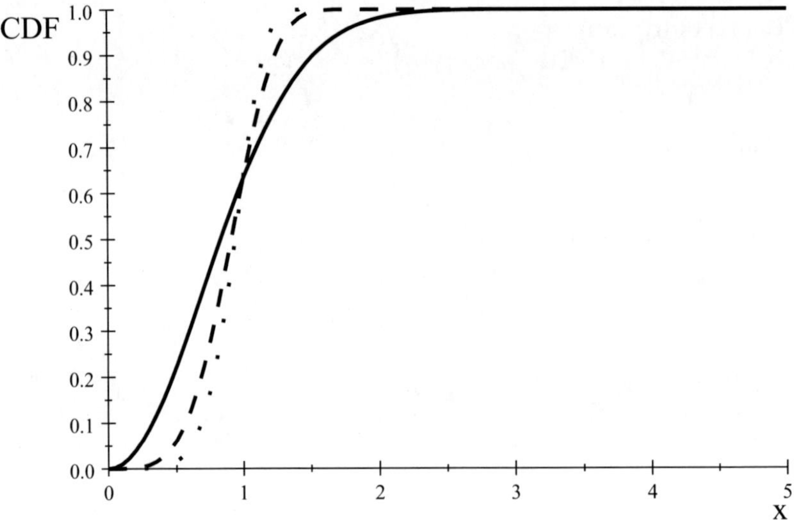

Figure 30.1.2. CDFs -WE (0,1,2)-solid, WE (0,1,4)—dash and WE (0,1,6)-dots.

30.2. Basic Properties

Mean $\quad \mu + \sigma\,\Gamma(1 + \frac{1}{\delta})$

Median $\quad \mu + \sigma(\ln 2)^{1/\delta}$

Variance $\quad \sigma^2[2]\Gamma(1+\frac{2}{\delta}) - (\Gamma(1+\frac{1}{\delta}))^2]$

Mode $\quad 0, \quad \delta \leq 1$

$\sigma(1-\frac{1}{\delta})^{\frac{1}{\delta}}, \delta \geq 1$

Moment about μ $E(X-\mu)^k$ $\sigma^k\Gamma(1+\frac{k}{\delta})$

Coefficient of variation $\dfrac{\sigma[2]\Gamma(1+\frac{2}{\delta})-(\Gamma(1+\frac{1}{\delta}))^2]^{1/2}}{\mu+\sigma\Gamma(1+\frac{1}{\delta})}$

Cdf $F_{(\mu,\sigma,\delta\}}(x)$ $1-e^{-(\frac{x-\mu}{\sigma})^{\delta}}$

30.3. Distributional Properties

Mgt $e^{t\mu}\sum_{k=0}^{\infty}\frac{(t\sigma)^k}{k!}\Gamma(1+\frac{k}{\delta})$

Cf $e^{it\mu}\sum_{k=0}^{\infty}\frac{(it\sigma)^k}{k!}\Gamma(1+\frac{k}{\delta})$

Hazard rate function $\frac{\delta(x-\mu)^{\delta-1}}{\sigma^{\delta}}$

The hazard function decreases for ð<1, constant for ð=1 and increase for ð>1.

$1-e^{-(\frac{X-\mu}{\sigma})^{\delta}}$ is distributed as uniform, U(0,1).

If $\delta=2$, then X is distributed as Rayleigh, $RA(\mu,\sigma)$.

If X is WE (0,1,1) then E (ln X) $=\gamma$, *the Euler constant*-and Var (ln X) $=\frac{\pi^2}{6}$.

Ln X is distributed as Gumbel distribution.

Beta Weibull distribution (BWE(a,b,c, λ)}

PDF $\dfrac{c\lambda^c}{B(a,b)}x^{c-1}\exp(-b(\lambda x)^c[1-\exp-(\lambda x)^c]^{a-1}$

$$x>0, a>0, b>0. c>0. \lambda>0$$

When a = 1 and b = 1, then BWE(1,1,c, λ) is WE (o,c,1/ λ)

CDF $$\frac{\Gamma(a+b)}{\Gamma(b)}\sum_{j=0}^{\infty}\frac{(-1)^j(1-e^{-(b+j)(\lambda x)^c}}{\Gamma(a-j)\Gamma(j+1)(b+j)}$$

30.4. Random Number Generation

Generate a uniform random variable U from uniform, U(0,1) distribution.

Set $X = \mu + \sigma(-\ln(1 - U) **(1/\delta)$.

Then X is a random variable from WE (μ, σ, δ) *distribution.*

Miscellaneous

Chapter M1

Levy Distribution ((LV(μ,σ))

Pdf f(x,μ,σ) $\sqrt{\frac{\sigma}{2\pi}}\frac{e^{\overline{2(x-\mu)}}}{(x-\mu)^{3/2}}$, x>$\mu$, $\sigma > 0$

The pdf of -LV*0,1), LV (0,3) and LV (0,5) are given in figure M1.1.

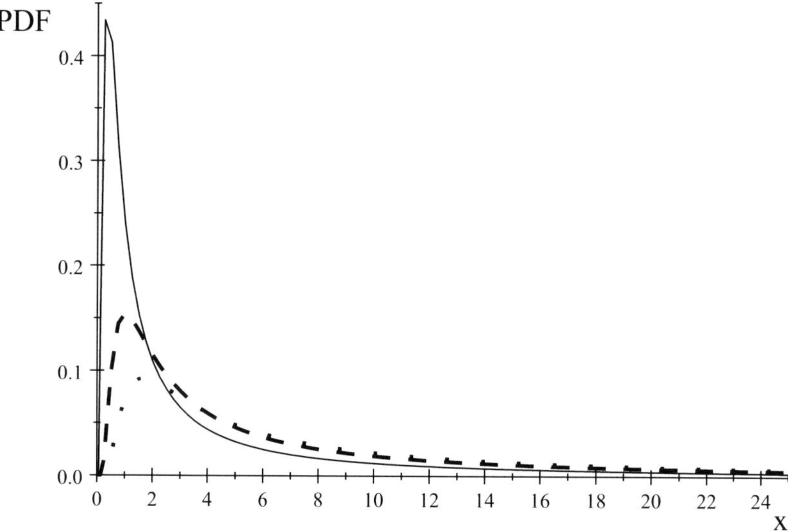

Figure M1.1. Pdfs LV (0,1)-solid, LV (0,3)-dash and LV (0,5) -dots.

CDF F(x,μ,σ) $2-2\Phi(\sqrt{\frac{\sigma}{x}})$

where $\Phi(x) = \int_{-\infty}^{x}\frac{1}{\sqrt{(2\pi)}}\,e^{-\frac{u^2}{2}}$ du.

The cdfs of LV (0,1), LV (0,3) and LV (0,5) are given in figure M1.2.

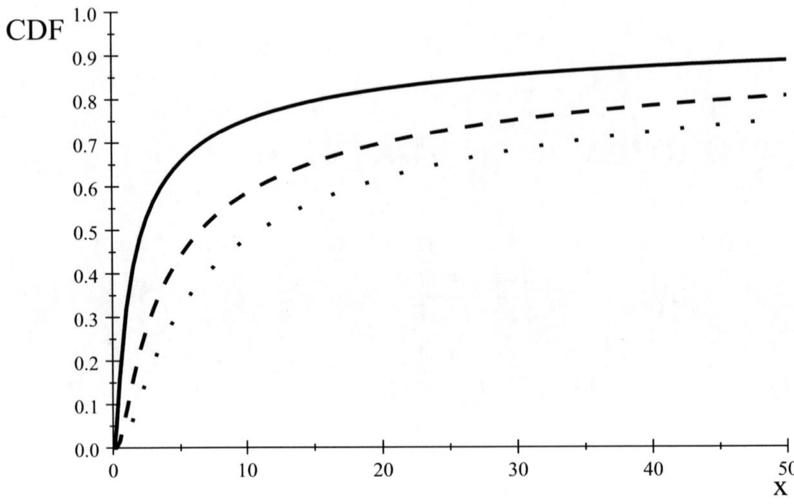

Figure M1.2. Cdfs -LV (0,1)-solid, LV (0,3)-dash and LV (5)-dots.

Mean ∞

Median $\mu + \sigma/2(erfc^{-1`}(\frac{1}{2}))^2$

Variance ∞

Moment generating function does not exist.

Characteristic function $e^{i\uparrow\mu t-\sqrt{}(-2i\sigma t)}$

If X has LV (μ, σ), the aX +b will have LV (k$\mu + b, k\sigma$)
If X has N (μ, σ^2) , the $(x-\mu)^{-2}$ will LV $(0,\frac{1}{\sigma^2})$

Chapter M2

Lomax Distribution (LM (α, σ))

Pdf $\qquad \dfrac{\alpha \mathrm{A}^{\alpha}}{(x+\mathrm{A})^{\alpha+1}}, \quad \mathrm{x} \geq 0, \alpha > 0, \lambda < 0.$

The pdfs of LM (1,1/2), LM (1,1) and LM (1,3) are given in figure M2.1.

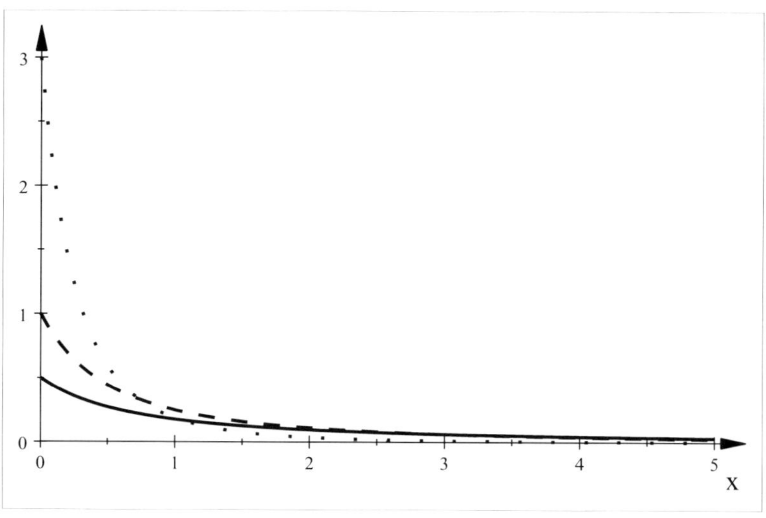

Figure M2.1. pdfs- LM (1,1/2)-solid, LM (1,1)-dash and LM (1,5) -dots.

Cdf $\qquad \dfrac{\alpha \lambda^{\alpha}}{(x+\lambda)^{\alpha+1}}$

The cdfs of LM (1,1/2)- LM (1,1) and LM (1,3) are given in figure M2.2.

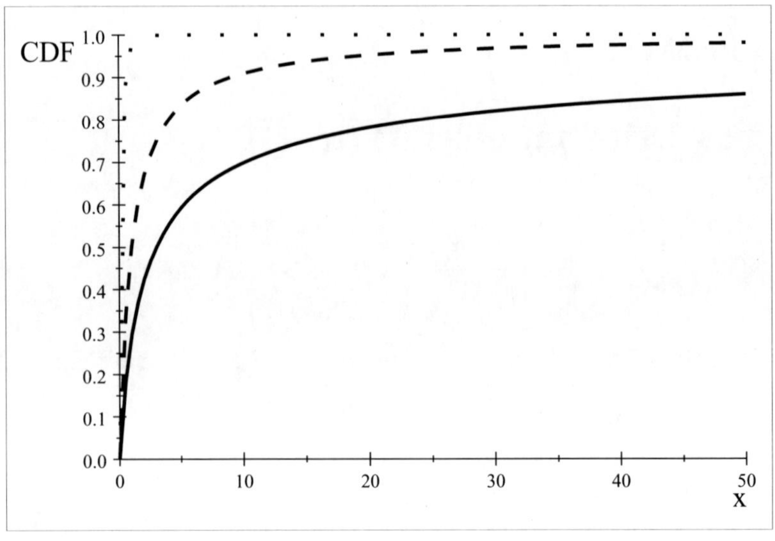

Figure M.2.2. Cdfs- LM (1,1/2)-solid, LM (1,1) = dash and LM (1,3).-dots.

Mean $\frac{\lambda}{\alpha-1}, \alpha > 1$

Median $(2^{\frac{1}{\alpha}} - 1)$

Moment generating function $\alpha e^{0\lambda}(-\alpha t)^{\alpha}\,\Gamma(-\alpha, \lambda t)$

Characteristic function $\alpha e^{i0\lambda}(-i\alpha t)^{\alpha}\,\Gamma(-\alpha, i\lambda t)$

Chapter M3

Triangular Distribution (TR(μ,σ))

Pdf

$$\frac{x-\mu}{\sigma^2} \quad \text{if } \mu \le x \le \mu+\sigma$$

$$\frac{\mu+2\sigma-x}{\sigma^2} \quad \text{if } \mu+\sigma \le x \le \mu+2\sigma$$

The following figure gives the pdf of TR (0,1)

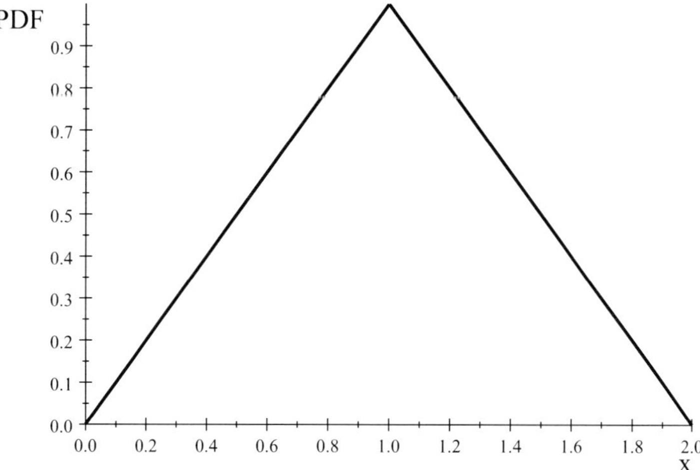

Figure M3.1. Pdf TR (0,1).

Cdf

$$\frac{(x-\mu)^2}{2\sigma^2} \quad \text{if } \mu \le x \le \mu+\sigma$$

$$1-\frac{(\mu+2\sigma-x)^2}{2\sigma^2} \quad \text{if } \mu+\sigma \le x \le \mu+2\sigma$$

$$1 \quad \text{if } x \ge \mu+2\sigma$$

The cdf of TR (0,1) is given in figure M3.2.

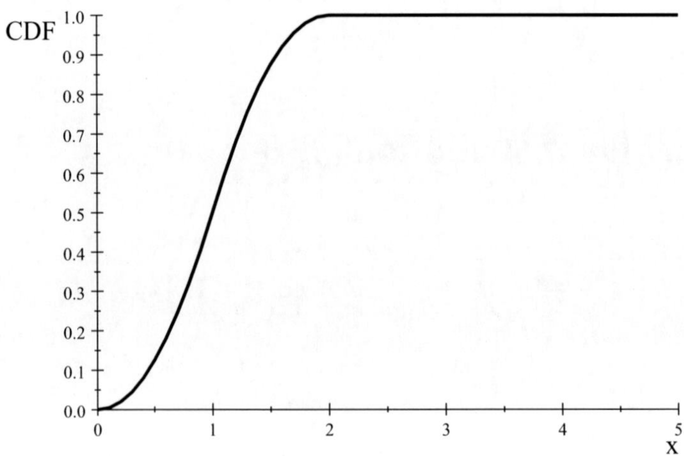

Figure M3.2. CDF- TR (0,1).

Mean $\mu+\sigma$

Mode $\mu+\sigma$

Variance $\frac{\sigma^2}{6}$

Random number generator

Generate two random variables U_1 and U_2 from, UN (0,1) distribution.

Set $X = +\sigma^2(\frac{U_1+U_2}{2})$. Then X is a random variable from, TR (μ,σ).

Chapter M4

Von Mises Distribution (VM (μ,σ))

Pdf $$\frac{e^{\sigma\cos(x-\mu)}}{2\pi I_0(\sigma)}, \mu-\pi \le x \le \mu-\pi, = \infty < \mu < \infty, \sigma \ge 0,$$

where $I_0(\sigma)$ is the Bessel function of the first kind.

The graph of VM (0,1) and VM (0,2) are s given in figure M4.1.

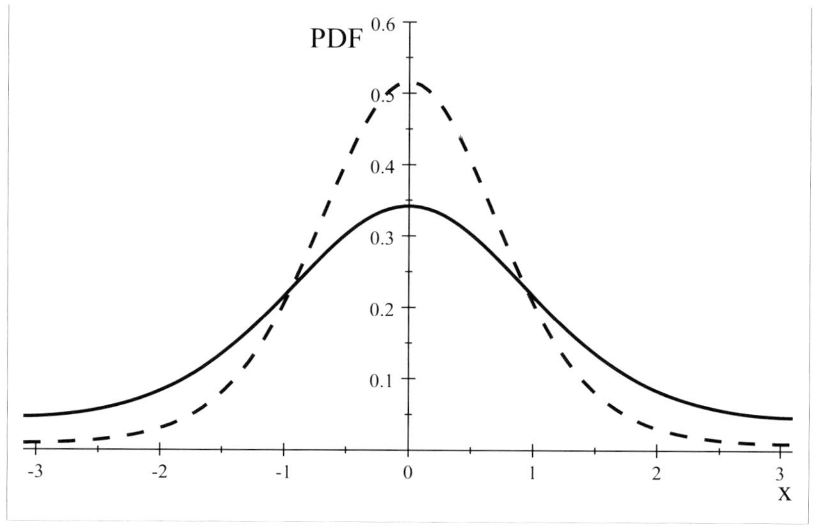

Figure M4.1. PDFS- VM ((),1)-solid and VM (0,2) -dash.

Cdf $$\frac{1}{2\pi}(x + \frac{2}{I_0(\sigma)}\sum_{j=1}^{\infty} I_j(\sigma)\frac{\sin j(x-\mu)}{j})$$

Mean μ

Median μ

Mode μ

Variance $1-\frac{I_1(\sigma)}{I_0(\sigma)}$, where $I_n(\sigma)=\frac{1}{\pi}\int_0^{\pi} e^{\sigma\cos x}\cos nx dx$

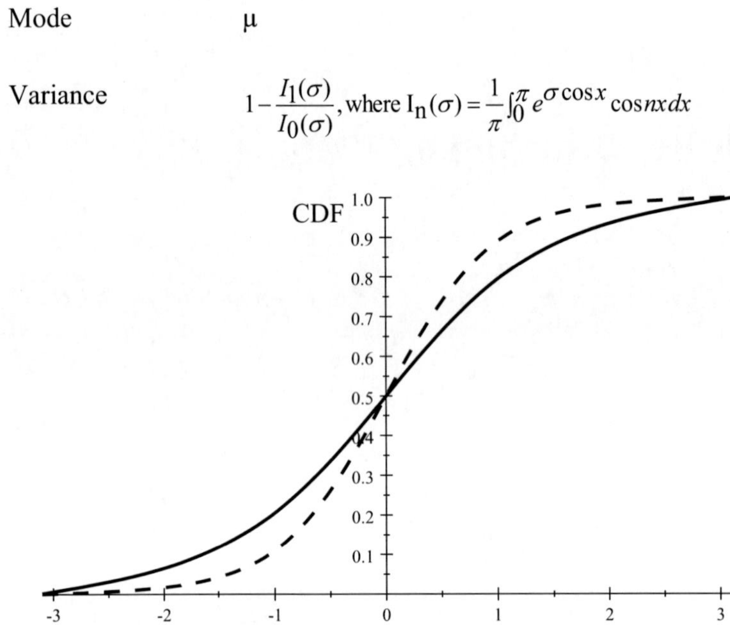

Figure M4.2. cdfs-VM (0,1)-solid and VM (0,2) -dash.

The pdf can be expressed as

$$\frac{1}{2\pi}(1+\frac{2}{I_0(\sigma)}\sum_{j=1}^{\infty} I_j(\sigma)\cos\{j(x-\mu)\}\text{'}$$

The circular moments of Von Mises distribution is defined as E(Z), where

z –e^{ix}.

The rth circular moment is

$$m_r=\int_\Gamma z^r f_{\mu,\sigma}(x)dx=e^{ir\mu}\frac{I_r(\sigma)}{I_0(\sigma)}$$

Jf σ=0, then VM(μ,0) is the uniform distribution in $(\mu-\pi,\mu+\pi)$.

Conclusion

In this book we have presented various univariate distributions and their basic properties. We also presented how random samples can be generated from these distributions. We hope the materials presented in the book will be very helpful for students and applied statisticians and further development in the research of these distributions.

References

Abramowitz, M. and Stegun, I. A. (1965). *Handbook of Mathematical Functions.* Dover Publications, New York, USA.

Ahsanullah, M. (1990). Some characteristic property of the normal Distribution. *Computational Statistics and Data Analysis.* 10, 117-120.

Ahsanullah, M. Kibria, B. M. Golam and Shakil, M. (2014). Normal and Student's t Distributions and Their Applications. *Atlantis Press in Probability and Statistics.* Paris, France.

Ahsanullah, M. and Hamedani, G.G. (2009) *Exponential Distribution Theory and Methods.* Nova Science Publishers. New York, USA.

Aitchison, J. and Brown, J. A. C. (1957) *The Lognormal Distribution*, Cambridge University Press, U.K.

Arnold, B. C. (1983). *Pareto distribution.* Fairland, MD, USA. International Cooperative Publishing House.

Balakrishnan, N. (Ed). (1992). *Handbook of The logistic Distributions.* Marcel Dekker New York, USA.

Blead, S. and Abdelal, A. (2018). Transmuted Arcsine distribution Properties and Applications, *Int. J. Res.* 10,1-11.

Bloch, *Daniel (1966).* "A note on the estimation of the Location parameters of the Cauchy distribution." *Journal of the American Statistical Association. **61** (316): 852–855.*

Bowman, K. O. *Shenton, L. R. (2007). "The beta distribution, moment method, Karl Pearson and R.A. Fisher" (PDF). Far East J. Theo. Stat. **23** (2): 133–164.*

Crow, E. L., and Shimiz, K. (eds.) (1988). Lognormal Distribution-Theory and Applications, Marcel Deker, Nyo.

Cane, Gwenda J. (1974). "Linear Estimation of Parameters of the Cauchy Distribution Based on Sample Quantiles." *Journal of the American Statistical Association*. 69 (345): 243–245.

Chattamvelli, Rajan., Shanmugam, Ramalingam (2021), "Laplace Distribution," *Continuous Distributions in Engineering and the Applied Sciences – Part II*, Cham: Springer International Publishing, pp. 189–199.

Chhikara, Rai S and Folks, J. Leroy. (1989). The Inverse Gaussian Distribution. *Theory, Methodology and Applications.* Marcel Dekker, New York, USA.

Crow, Edwin L., Shimizu, Kunio, eds. (1988), Lognormal Distributions, Theory and Applications, Statistics: *Textbooks and Monographs,* vol. 88, New York: Marcel Dekker, Inc., pp. xvi+387.

Dubey, Satya D. (December 1970). "Compound gamma, beta and F distributions." *Metrika*. **16**: 27–31.

Folks, J. Leroy, Chhikara, Raj S. (1978), "The Inverse Gaussian Distribution and Its Statistical Application—A Review," *Journal of the Royal Statistical Society,* Series B (MethodoLogical), 40.

Gupta, Arjun K., ed. (2004). *Handbook of Beta Distribution and Its Applications*. CRC Press, Boca Raton, Florida, USA.

Haight, Frank A. (1967). *Handbook of the Poisson Distribution.* New York, NY, US: John Wiley & Sons. ISBN 978-0-471-33932-8.

Hirseh. Werner Z. (1957). "Binomial Distribution—Success or Failure, How Likely Are they?." *Introduction to Modern Statistics.* New York: MacMillan. pp. 140–153.

Haight, Frank A. (1957). *Handbook of the Poisson Distribution.* John Wiley and Sons. New York, USA.

Hallinan, A. J. (1993). A review of Weibull distribution. *Journal of Quality Technology, 25*(2), 85–93.

Heyde, CC. (2010), "On a Property of the Lognormal Distribution," *Journal of the Royal Statistical Society, Series B*, vol. 25, no. 2, pp. 392–393.

John S. deCani & Robert A. Stine (1986). "A note on deriving the information matrix for a Logistic distribution. *American Statistical Association:* 220–222.

Johnson, N. L. and Kotz, S.(1970). *Distributions in Statistics: Continuous Univariate Distributions:* Vol 1. John Wiley & Sons, New York, USA.

Johnson, N. L. Kotz, S., Kemp, A. (1993). *Univariate Discrete Distributions* (2nd ed.). John Wiley and Sons., New York, USA.

Lancaster, H.O. (1969), *The Chi-squared Distribution*, Wiley and Sons. New York, USA.

Laplace, P-S. (1774). Mémoire sur la probabilité des causes par les évènements. *Mémoires de l'Academie Royale des Sciences Presentés par Divers Savan,* 6, 621–656.

Patel. J. K. and Read, C.B. (1997), *Handbook of the Normal Distribution.* Marcel Dekker, New York, USA.

Phillips, P. C. B. (1982) "The true characteristic function of the F distribution," *Biometrika*, 69: 261–264.

Prochaska, B. J. (1973). A note on the relationship between geometric and the exponential distributions. *The American Statistician,* 27, 27.

Rothenberg, Thomas J., Fisher, Franklin, M., Tilanus, C.B. (1964). "A note on estimation from a Cauchy sample." *Journal of the American Statistical Association.* **59** (306): 460–463.

Sheshadri, V. (1998). *The Inverse Gaussian Distribution. Lecture Note in Statistics,* No.137. Springer Verlag, New York, USA.

Tadikamalla, Pandu R. (1980), "A Look at the Burr and Related Distributions," *International Statistical Review,* 48 (3), *337*–344.

Tweedie, M. C. K. (1957). "Statistical Properties of Inverse Gaussian Distributions I." *Annals of Mathematical Statistics. 28 (2,: 362–377.*

Index

H

I

L

M

N

O

P

Q

R

S

T

U

V